Speaking for Ourselves

Edited by Julian Agyeman, Peter Cole,
Randolph Haluza-DeLay, and Pat O'Riley

Speaking for Ourselves
Environmental Justice in Canada

UBCPress · Vancouver · Toronto

17 16 15 14 13 12 11 10   5 4 3 2

Printed in Canada on acid-free paper ∞

Library and Archives Canada Cataloguing in Publication

Speaking for ourselves : environmental justice in Canada / edited by Julian Agyeman . . . [et al.].

Includes bibliographical references and index.
ISBN 978-0-7748-1618-2

1. Environmental justice – Canada. I. Agyeman, Julian

GE240.C3S64 2009   363.7'00971   C2009-900924-2

Canadä

UBC Press gratefully acknowledges the financial support for our publishing program of the Government of Canada through the Book Publishing Industry Development Program (BPIDP), and of the Canada Council for the Arts, and the British Columbia Arts Council.

This book has been published with the help of a grant from the Canadian Federation for the Humanities and Social Sciences, through the Aid to Scholarly Publications Programme, using funds provided by the Social Sciences and Humanities Research Council of Canada.

UBC Press
The University of British Columbia
2029 West Mall
Vancouver, BC V6T 1Z2
604-822-5959 / Fax: 604-822-6083
www.ubcpress.ca

Contents

Acknowledgments

Julian would like to dedicate this book to those people in Canada and around the world who have fought for, and continue to fight for, a more just and sustainable world.

Peter would like to acknowledge his family, including Suzanne Cole and Shane Neufeld, and friends for their support and love, as well as all beings of spirit struggling alone or in solidarity within spaces and times of pain, fear, triumph, celebration, awareness, and compassion. We are the land and we are all related.

Randy would like to acknowledge students at The King's University College and his children, Taryn and Ethan Haluza-DeLay – all of whom give energy and hope and are among those for whom a just, sustainable land must be their future.

Pat would like to dedicate this book to the ancestors who have cared for this land since time immemorial; to her children and their families – Lee O'Riley and Joey Dingwell, Chris, Kiwa, Michio, and Koji Babcock, Lori, Ted, Amber and Jessica Aston; to her sisters, Kym Wark and Lorraine Mackie; and to those yet to come.

Collectively, we would like to acknowledge the help we have had in creating this book: the many people who have worked on portions of the book or whose efforts have been part of the struggles represented in these chapters. They are too many to list, and we hope that the coming years will see far more people engaged in working for environmental justice. We also wish to acknowledge Randy Schmidt, editor at UBC Press, who encouraged us despite our perhaps overly idealistic approach, and helped bring this book to fruition.

Prologue. Notes from Prison: Protecting Algonquin Lands from Uranium Mining

Robert Lovelace

For a monarch, dictator, or democrat, the dilemma of ruling is this: if rule is accommodating, then the ruled become indolent and ungrateful; if the ruler is harsh, then the ruled become uncooperative and rebellious. In such a paradigm the only way to maintain power is to keep subjects off balance and marginalized through seduction and force. This dilemma in governance arises out of a heritage of colonialism; that is, the process by which a group having exhausted its sustainability options dispossesses another group of its. Using seduction or force, a dominant group undermines the power of multiple others. This is the principle mechanism that separates much of humanity from the sacred relationships with the earth and has become normalized in the governance of nations.

No matter how present-day leaders disguise this process of resolving resource "development," the apparent consequence is a global totalitarian catastrophe in which all possible human resources (and those of most other species) are consumed while the indigenous knowledge to survive in the real world is lost. Determinism, progress, and ideological or theological perfectionism are dead-end illusions on a living planet in this galaxy. Here, the highest achievement for humankind is to exercise our innate ability to adapt to the local. In knowing and using the local environment sustainably, the dilemma is expunged and each individual takes a transactional responsibility, and experiences immediate consequences. Collectives are supported for what individuals give by way of sustenance and knowledge, not just what they can take away for less than they contribute. The question is then, how can humankind achieve such a state of grace.

I am writing this prologue from the Central East Correctional Centre in Lindsay, Ontario. The name of the institution is misleading. There is little attempt made at corrections here, save the fact that some people believe denying basic human needs and humiliating inmates will in some way change the response to the same kind of alienation and poverty that are the antecedents that bring them in conflict with the law in the first place.

This is a modern "super maximum security" prison designed specifically to house the greatest number of inmates while using the least amount of labour and capital. This efficiency is best illustrated by the fact that although the prison's worship centre was dedicated five years ago, it has never been used. I have now served seventy-one days of a six-month sentence. I spend sixteen and a half hours a day locked in a two-metre-by-four-metre cell. The overhead light is never turned off. When out of my cell I share a common area with thirty-two other men who use their best instincts to pass the time and get along. This prologue was originally composed using a four-inch-long pencil that had to be sharpened on the abrasive tread of the stairs leading to the second tier of prison cells. Almost everything of value, whether for recreation, communication, or survival, has some component of local invention. The diversity of creation, the alertness of mind, and the indomitable human spirit remains alive even within these prison walls.

I was imprisoned for refusing to purge contempt of an Ontario Superior Court injunction that ordered my community, the Ardoch Algonquin First Nation, to abandon a secured entrance protecting an area of our homeland from uranium exploration. In the winter of 2006-2007, a mining exploration company, Frontenac Ventures, had without official notice staked thirty thousand acres, north, east, and south of Crotch Lake, and occupied an old tremelite-processing plant near Robertsville. The Ontario Ministry of Northern Development and Mines (MNDM) registered these claims and the Ministry of Natural Resources (MNR) provided leases for much of the land not privately owned. The Algonquin homeland and Ontario's claim to jurisdiction overlap in eastern Ontario without any historical treaty or land cession. When Ontario acted unilaterally, without consulting our community, the Province broke the law. Consultation is required before development is contemplated where jurisdictions are shared or in dispute. Along with consultation there must be comprehensive and reasoned accommodation of Aboriginal rights and cultural needs. These principles have been confirmed by the Supreme Court of Canada as necessary in preserving the honour of the Crown in relations with Aboriginal communities and nations. The historical practice in Ontario has been to refute Aboriginal claims and address individual issues on a case-by-case basis rather than develop practical policy. The denial of appropriate levels of responsibility has been Ontario's consistent course of action in meeting constitutional obligations to Aboriginal Canadians. The injunction arose out of this confusion.

When our community protested to Ontario in May of 2007 that staking had taken place and extensive deep-core drilling was imminent, MNDM consented to a meeting. The meeting was attended by six members of the Ontario Provincial Police (OPP), representatives of Ardoch and Shabot

Members of Ardoch Algonquin First Nation and of the area community protest in 2007 against uranium exploration and mining on Algonquin lands. Photo by Susan DeLisle.

Obaadjiwan First Nations, and a low-level MNDM geologist. The geologist threw half a dozen technical maps on the table, said there was nothing to fear about exploration, and called that consultation. The OPP wanted to know if they should expect conflict to arise. We assured the OPP that we had no intentions to take violent action but that we would use a four-pronged approach of: research, community education, legal action, and direct action in a strategy to protect our interests should Ontario not respond more appropriately.

In mid-June 2007, having heard nothing more from Ontario, the Ardoch Council requested Frontenac Ventures remove its personnel and equipment from Algonquin land. The company complied with this directive and on 28 June Ardoch and neighbouring Shabot Obaadjiwan First Nation set up a security barrier at the abandoned Robertsville ore-processing plant. The OPP established a security zone of their own in an effort to minimize conflict. Over the next month, the security barrier developed into a camp, at times with a population of forty residents.

Ontario remained silent. Frontenac Ventures attempted to divide the First Nations by promising Shabot Obaadjiwan money and benefits of development. When this failed, the mining company offered both First Nations $10,000, as well as promises of some employment and an archaeological study that would protect burial sites. Ardoch and Shabot refused to take the bait. Two principles were well entrenched for the First Nations: the Crown's duty to consult, and a strong sense of Algonquin values. Both communities recognized that Ontario's duty to consult had not been met. At best, Ontario stated that consultation and accommodations were to be addressed through the Algonquins of Ontario Land Claim, although protection of land from mineral staking had never been raised during fifteen years of dysfunctional negotiations and Ardoch had withdrawn with misgivings from the process some six years earlier. The First Nations knew that the Supreme Court decisions regarding Aboriginal rights and title were clear in directing Canadian governments to consult directly with affected communities. They were not about to surrender what other communities across Canada had fought hard to achieve. If Ontario would not protect Aboriginal rights, then they felt obliged to do it for themselves.

Second, although Shabot and Ardoch are politically separate, they represent a large interdependent and interrelated community with overlapping original jurisdictions covering three watersheds. There is a strong inclination toward tradition in both communities. Although a century of integration has eroded old-time labours and cultural practices, there is still a fierce adherence to the ideals of Algonquin life. Ever since Ardoch fought a twenty-seven-day standoff in 1981 to protect wild rice at Mud Lake from government encroachment, people are diligent in protecting expressions of Algonquin culture that remain. There was no doubt that Ardoch would take a militant stand in protecting the land.

To address the threat of uranium exploration and mining, the leadership of the Ardoch community adopted our four-pronged strategy of research, community education, legal action, and direct action. In concert, these form a dynamic plan. The task of leadership is to coordinate them for efficiency, conservation of resources, and overall impact. Ardoch understood that resistance was not enough. The four-pronged approach was needed to achieve a successful outcome, one in which not only exploration for uranium would be stopped but also existing legal precedents would be strengthened and potentially the long-term struggle for acceptance of Aboriginal rights and title would be advanced.

The four-pronged approach is not new. It emerges organically from grassroots struggles in defence of community interests. I first heard it articulated among the Cree of the Nelson River in Manitoba. Expressing the strategy logically allows for measurement of process and outcomes, but the force of the plan is also found in imaginative interaction. Almost always there is something among the four activities that individual community members can identify in their own pool of talents. In such a dynamic, leadership needs to be transactional among all participants. A too-structured leadership model will have tendencies to overlook important talents or assign extremely talented people to tasks that they may be ill-equipped to perform. Transactional leadership allows for the emergent forces of a movement to be activated with any one person or group assuming leadership when real situations demand. This is a traditional form of Anishnàbèk leadership. Of course, there are inherent difficulties with such a system, but practice and knowledge can lead to consistent success. There are people who are accustomed to formal structure. Initially, they can feel lost without very clear lines of "responsibility." There are others who believe they have a right to leadership or that the ideology or strategy they espouse is the only way to go. The shift in paradigms can be tricky. For indigenous communities that have traditional social values, it is not so hard. Some NGOs, settler groups, and indigenous groups long dominated by colonial control through instruments such as the Indian Act have much more difficulty adjusting to situational or transactional leadership. The benefit is that everyone can participate in learning, teaching, and leadership, and leadership as a function becomes better understood and accepted as feelings of mutual respect and benefit develop.

Research is the engine of intelligence, and intelligence is the foundation of a successful movement. The old saying about the strength of an idea whose time has come is true, but the idea must be honest, accurate, and understandable. As a small indigenous community, Ardoch did not know much about uranium. Our first concerns were with a trapline and the failure of Ontario to consult. Research at the beginning stages allowed our community to expand its understanding of the history of uranium mining and the associated social, economic, and environmental consequences. We became knowledgeable and then capable of linking our concerns with the broader world. As well as expanding our knowledge outward, we opened the bundle of traditional knowledge and found within the many coded layers of direct reference to our current situation and the language needed to express our concerns in universal terms. Traditional knowledge illuminates a vision of the way forward that strengthens courage and insight.

Research should also be used to understand the nature of opposition. In today's world it is easy for people to cover their tracks. As part of our intelligence efforts we like to know "who's who." We study academic work, career histories, and photographs in the attempt to understand why proponents and bureaucrats do what they do.

Community education is part of the four-pronged strategy. Action follows enlightenment, and enlightenment comes from knowledge. Community education combines both knowledge and action. Sharing knowledge with the largest number of people is essential to win public opinion and advance long-term change in collective consciousness. A full-page advertisement in a leading daily newspaper can be impressive as well as costly. However, the best community education takes place around the kitchen table between friends and relatives. The best community education begins on the inside, expanding from there. A knowledgeable community has the capacity to educate others. Ardoch began our community education at a council meeting. The entire council, community members, and representatives of the neighbouring First Nation viewed and discussed the National Film Board documentary *Uranium*. The film is available in libraries and for rent on DVD. Soon the film was being watched in living rooms, community halls, and on personal computers. Guest speakers were joining many community forums. A proliferation of information about uranium mining began to grow as people became hungry for more knowledge. People generally felt comfortable engaging in the discourse no matter what their level of education or walk of life. Men and women shared their perspectives from an informed point of view. Sharing knowledge became a way to heal old wounds, rekindle friendships, and unite the community and the larger concerned public. Results were measurable. The first public meeting at the community hall in Sharbot Lake had drawn only three people. A year later, many thousands of people know about uranium exploration and mining and have voiced their concerns about protecting the environment and Aboriginal interests.

Canadian governments are divided into three separate legal orders. Parliament and legislatures create law, cabinets administer law, and the judiciary decides how laws can be applied. In the four-pronged approach, we take all of these into consideration when engaged in legal action. Contrary to popular opinion, the courts are not very useful in deciding issues relating to Aboriginal people. Poor decisions effect subsequent rulings, building layers of harmful precedent into a tide of history. It can take decades to remedy inadequate decisions made by ill-informed judges. The court process is extremely costly and accentuates the disparity between indigenous communities and colonial governments. Canadian jurisprudence is historically prejudiced in favour of colonial values and aspirations. Finally, although Canadian law has its foundations in the doctrine of continuity in regard to the Aboriginal right of self-governance, Canadian courts are restricted by letter and politics to acknowledge only colonial law. Aboriginal people who choose to defend their own laws must hold Canadian courts in contempt for their failure to uphold natural justice. The dual pretence of fairness and faithfulness to the rule of law makes Canadian courts laughable if not for the serious harm that is perpetuated on Aboriginal people. Lawyers who represent Aboriginal people are caught in the dilemma of trying to provide the best possible representation before courts that deny the validity of their clients' customary laws and culture.

In the absence of the administration of existing laws that protect Aboriginal rights and title, indigenous communities are vulnerable to corporate agendas and opposing political interests. This is what has happened in our situation. When Ardoch demanded a previously defined right to consultation, MNDM sloughed off its responsibility. Ardoch was forced to choose the unilateral action of blocking Frontenac Ventures' activities. The corporation countered by suing Ardoch and its leadership for $77 million. At court, Frontenac Ventures obtained an injunction ordering Ardoch to abandon its security barrier and permit extensive deep-core drilling. Before the court, the Government of Ontario muddled the issue of consultation by refuting its responsibility to "unorganized" Aboriginal communities, and the uninformed judge was reluctant to make substantial constitutional decisions in the early stages of the proceedings. Ontario's tactic of deny and stall was successful. Ardoch refused to respect the injunction and continued to maintain the security barrier. In response the court refused to recognize Ardoch's lawyer and eventually found three of Ardoch's leaders in contempt of court. While Frontenac Ventures prosecuted the contempt charges, the Government of Ontario remained silent about its legal duty. In cross-examination, Ontario's lawyer accused the Ardoch leadership of constructing its own "self-help" remedy.

Both the government and the corporation were able to manipulate the court process into minimizing the ability of the First Nation to protect itself and avoid their own liabilities. With an emphasis on Ardoch as a "law breaker," they avoided close scrutiny of their own obligations. In reality, the Government of Ontario was liable to Frontenac Ventures for issuing permits and registering claims before meeting its legal duty to Ardoch, and Frontenac Ventures had proceeded with its activities even after it was aware of the Government of Ontario's obligations. The court ruling on contempt charges reinforced colonial values that hold corporations and governments above clear constitutional commitments. A neutral observer might have suggested that Ardoch have trusted more in the court for a fair and unbiased hearing before choosing to ignore the injunction issued by Judge Thompson. However, neutrality never existed. During the first days' hearings when Shabot Obaadjiwan's lawyer raised the Royal Proclamation of 1763 as a guide to understanding title issues, the judge had waved his arm and pronounced that he was not interested in the past, only the "here and now."

Challenging the tide of history in Ontario courts results more often than not in long and costly litigation that frustrates justice. When defence in court fails or bogs down, legal action needs to shift weight to the arena of law in practice. This is the domain of politicians. Understanding the principles and history of legislation is of prime importance in shifting policy on the application of laws. In Ontario, environmental protection and the ideals of environmental justice are weak. Legislation finds its substance more in policy than in the letter of the law. In Ontario, even the quasi-legal tribunal decisions are subject to the interpretive vagaries of partisan politics. Even less substantial is the Government of Ontario's commitment to Aboriginal justice. To their credit, many Ontario ministries have developed policies of inclusion and accommodation for Aboriginal people; however, there is always a caveat that such policies are not an affirmation of an Aboriginal right. The MNDM has no policy at all and benefits from legislation minted at a time when Aboriginal people were expected to be erased from the Canadian landscape. MNDM has no written protocol or directives that instruct field technicians or administrators on provincial obligations to Aboriginal people. MNDM enjoys a closed constituency. The mining industry is its client. The Government of Ontario receives revenue from mining with almost no immediate reciprocal costs. Internationally (and this includes Canada), the mining industry is able to externalize its overhead by degenerating environments and impoverishing indigenous cultures. Not only does mining operate within a set of laws that are outside common propriety, the regulations governing the industry promote speculation for profit based on the ability to escape human and environmental liability. Although politicians and law makers can enjoy the profitability of mining, they are also trapped in a quandary of promoting one set of democratic principles for the general population and another very different set of principles based on unearned privilege for the mining industry. This is certainly a call for action in all three orders of Canadian law and represents the political vulnerability that gets people thinking.

Among traditional Algonquin healers it is said that a single plant's medicinal properties are strengthened when combined with other complementary medicines. At the beginning of this prologue I mention four prongs: research, community education, legal action, and direct action. I say that in concert they become a formidable strategy. Although navigating the Canadian legal system seems treacherous for Aboriginal peoples, when supported by research, community education, and direct action, legal remedies can be achieved through persistence. Court rulings though are often only peripheral to law in practice and strong legal activity can produce substantial progress through negotiations. Among forward-thinking Canadian politicians there is a movement toward establishing accords that are legally binding, respectful of cultural differences, and reviewable. Through this practice, the experience of mutual respect and benefit through the life of an accord can lead to substantial changes in legislation.

Shifting colonial ideology from its present foundations in Canadian law requires much more than changing the words. Bill-C31 is a good example of how legal redefinition can undermine the strength of indigenous identity in the name of human rights and give colonial powers just another avenue to terminate rights and title. Action, doing, becoming, growing –

these are the vital processes that bring about lasting change. Such processes of change cannot be imposed but rather must emerge at a local level and represent the innate principles of the living cultures from which they find ascendancy. This is the power of direct action. Often mistaken as impromptu, violent exhaustion, or last-resort anger, direct action is able to draw attention to a cause. However, to be effective, direct action must not be an end in itself for the attention, then it works against those who use it. Direct action must be considered, planned, purposeful, and controlled. Energized through evolving forces, direct action is best channelled through collective self-discipline, self-awareness, and clear objectives. Like other initiatives in the four-pronged strategy, direct action should be integrated throughout all activities. It is the primary tool for building momentum and often the signature of the underlying philosophy of a movement.

Idleness in a population is preferred by leaders who subscribe to privilege and by commercial interests that depend on political acquiescence. Both find populations that begin to think and act on their own interests as threats. Rather than using outright suppression, Western governments and markets have learned to co-opt emergent ideas and energy. Direct action with a philosophical core becomes at once a threat to them and an opportunity. Rebranding direct action becomes their strategy for deflecting honest criticism. For a variety of reasons, Aboriginal people's direct action more often than not has taken on the characteristics of resistance. For their own reasons, media enhances the image of Aboriginal activists as radical, savage, and warlike. Some Aboriginal activists have adopted this image as a sure way of drawing attention to their cause. However, as indigenous people who must oppose colonialism to protect our homelands and way of life, we run the risk of becoming largely cultures of resistance or, worse, perpetual victims. The very essence of living in harmony with the environment, sharing with others, and reducing conflict is undermined by persistent defence of these values. Because of this, strategic direct action should be mindful not only to project indigenous values but to enliven them as well. Indigenous activists need to stay close to the land and community to draw continuing insight and energy, and to reinforce traditional attitudes and beliefs. Direct action should take its shape and purposes from the intrinsic goodness embedded in indigenous epistemologies.

Environmental and Aboriginal justice converge on many levels and often share ground during direct-action events. At the heart of the injustices that these groups hope to eliminate are the tenets of colonialism. Colonialism supports the Canadian dominion as a system that claims the privilege of pillaging the earth and displacing the original human beings for its own wealth and security. It is accepted that Canadian citizens will benefit from such a system. Environmentalists and indigenous people understand that this kind of privilege will eventually destroy real wealth and make living on this earth intolerable. In the struggle for justice there are sometimes issues that divide natural allies. Individual and collective interpretations and interests will arbitrarily separate goals. The very process of adapting locally has the tendency to create parochial attitudes and loyalties. Direct action is a positive way to unite people and educate communities around shared principles. As trite as it may sound, working together, sharing responsibility, and having fun together are the best antidotes for the misery of separation and internal violence. Direct action in its finest form is about discovery, trial and error, imagination, and celebration. Peace is not a goal, it is a process.

In times past, indigenous cultures tended not to be centred on material wealth. Certainly, personal belongings were meaningful. Technological devices and goods made survival surer and adaptation easier. But these were largely knowledge-based cultures. Understanding indigenous languages tells us this. Our indigenous knowledge systems are complex, holding generations of information about land, ecological processes, human relations, and spiritual realms. Our cultures were resilient and resourceful because individuals had access to the collective knowledge and wisdom of their family, clan, and people. Communities computed their own needs and portioned out resources with a sustainable logic that insured an equitable share for everyone. Equity was not a moral question; it was an assurance of mutual benefit and protection. While our indigenous cultures were uniquely generous and complete, they were not invincible. Their very ability to adapt when confronted with western European economic hegemony may well have been a prime cause in their eventual submission. Today we live with the consequences – impoverished and confined with our resources controlled by foreign powers, corporations, or governments. Resistance by opposition alone cannot succeed in preventing our final destruction. Stepping out of the shadow of the *Windìgò* and becoming one with *Windìgò* is not the answer either. Communities that survive will be those that reverse the loss of traditional knowledge and belief by reinvigorating individual talents and collective interests in living again in real relationship with the land. For it is the land that shapes us. It is the land that gives us our name. It is the land that defines our boundaries. We begin by taking care of our Elders and taking back our Children.

It is easy to identify a threat against Aboriginal people when it takes the shape of a company exploring for uranium. And in many ways, defeating such a threat is easier than overcoming substance abuse, addictions, corruption, and greed. But as surely as it will take restoring the land to health, to return to creation, people must restore themselves. Aboriginal people cannot turn to money or foreign governments to do it for them. The four-pronged strategy and transactional leadership are key tools for fighting exploitation, but they can also be used to restore our physical, emotional, intellectual, and spiritual integrity. Warriors must think and act beyond the barricades to be really effective. Warriors who build community and culture, protect the young and old, and strive for peace with their neighbours will bring about positive growth in our communities.

As long as democracy and colonialism can exist hand in hand, human life for the vast majority of human beings will be a misery. Even when fighting against colonialism and in favour of environmental protection, the costs are high. Although individual victories are cause for celebration, the price in community resources is extensive. Fractures in social relationships and the traumatic effects of isolation and abuse and exposure to political injustice remain long after success. Community healing needs to be a core principle for education and action in communities that engage in the struggle to restore their independence and culture. The warriors' responsibility is to encourage and protect the healing process in their communities.

I have come to understand that the struggle for environmental justice is rooted in the hearts of people of all races and cultures. It finds its original impulse in the inherent sense of indigeneity that cannot be extinguished. For Aboriginal people in Canada, that impulse is still raw, a scar that has yet to close. Because of this we are at once a symbol and lamp in the battle to save the planet. We still hold legal tenure to the land. Our knowledge systems, although compromised, continue to form the structure and hold a vast content of information that supports sustainable living. Our epistemologies continue to bear witness to the goodness of humanity and the love of a Creator and beautiful Creation. As I sit here writing in the early morning in my cell I know that my friends and neighbours are awakening to the task at hand. I am humbled and assured that the tide of change is rising.

On 28 May 2008, Robert Lovelace was released from prison after a successful appeal of his sentence. He had served 104 days of a six-month sentence for peaceful protest. The written appeal decision set important precedents for courts in dealing with political dissent and the inclusion of customary Aboriginal law in the Canadian legal system. Frontenac Ventures Corporation filed a leave to appeal the Ontario Appellate Court decision to the Supreme Court. The Supreme Court upheld the Ontario Appellate Court decision on 4 December 2008 and awarded costs to the Ardoch Algonquin First Nation.

Speaking for Ourselves

Introduction. Speaking for Ourselves, Speaking Together: Environmental Justice in Canada

Randolph Haluza-DeLay, Pat O'Riley, Peter Cole, and Julian Agyeman

Images of First Nations canoes entering New York City Harbor were newsworthy around the world in 1991. Grand Chief of the Northern Quebec Cree, and later national chief of the Assembly of First Nations, Matthew Coon Come led James Bay Cree and Inuit paddlers from Quebec, starting at James Bay, through Ontario and down the Hudson River. The voyage was to protest the State of New York's agreement to purchase electricity from the Province of Quebec, which had dammed the rivers of their homeland and built hydroelectric projects on unceded lands without consultation or permission from the James Bay Cree. The massive hydroelectric project – known as the James Bay Project because it had primarily dammed or diverted rivers that flowed into this large arm of Hudson's Bay – covered an area the size of the state of New York. Construction had begun in 1971, with no notification to those who would be most affected, including non-human forms of life. It is no exaggeration to argue that the invisibilization of Aboriginal peoples in Canada as contemporary stewards and keepers of the land is a consistent tendency (if not policy) within the Canadian cultural imaginary as co-constructed by government, media, and various transnational corporate entities.

Between 1971 and 1990, the effects of earlier hydroelectric projects included reservoirs and other kinds of water diversions that flooded and fouled huge tracts of lands and waterways used by Aboriginal peoples for hunting and trapping and gathering, while also negatively impacting the animal and plant nations. Meanwhile, permanent new roads connected Cree and Inuit communities with the south, bringing great social change to formerly isolated sustainable communities. Cultural contact with the south, together with untoward environmental changes, disrupted local economies, replacing sustainable subsistence with poverty, welfare, and social breakdown. Few Cree people were employed by construction or other infrastructure operations of local economic developments.

The first James Bay Project announcement catalyzed Aboriginal activism in Quebec and across the country. By 1973, a court decision had forced the Quebec government to negotiate agreements with Native peoples in the affected lands. The James Bay and Northern Quebec Agreement of 1975 became one of the first "modern" Aboriginal land claim settlements in Canada. This project was only the first stage of extensive hydroelectric diversion of waterways in northern Quebec. The second phase of the project received even more vocal protest, especially after the Great Whale River project was proposed. The canoe protest in 1991 was part of this campaign (Grand Council of the Crees, 1998).

When the State of New York cancelled its plans to purchase more hydroelectric power from Quebec – citing concerns over Native rights and environmental impacts – construction was halted. A new arrangement had to be made with the First Peoples of northern Quebec. In 2002, the Cree and the Government of Quebec signed the landmark Paix des Braves, or Agreement Concerning a New Relationship. It is widely described as a nation-to-nation agreement, a far cry from the invisibilization of a people who just thirty years before had been ignored and treated as anachronisms, holdovers from prehistoric or at least premodern times to be swept aside by the hubris and dismissal of modern Canadian government-industrial development partnerships.[1]

This case represents important processes in Canadian history, and in environmental and social analysis, bringing together many conversations and many ways of knowing and being. Like many other Canadian cases, the James Bay Project has not been perceived to be one of environmental justice but one of racism as practised by the state and its corporate institutional partners. The James Bay fiasco has also been seen to exemplify attitudes and practices of ongoing colonization of First Peoples, and dismissal of the other forms of nonhuman life on the land and in the waters and skies. Yet, while land and energy are usually considered environmental issues, the account of Aboriginal peoples in Canada is usually described as social injustice.

In contrast to the visibility of the environmental justice movement in the United States, Canada's environmental justice movement has been labelled and contextualized within the purview of social justice and human rights advocacy movements. There *have been* environmental justice movements in Canada for centuries (if not millennia), just as there have been struggles over the ownership and use of land and resources throughout Canadian history – and before the boats came. The struggles of Aboriginal peoples over the loss and destruction of their traditional homelands, as well as of their traditional ways (including economies) of living, are clear examples of land and freedom movements. A number of scholars argue in this book that advocacy in Canada for what is called "environmental justice" has taken different forms (see also MacGregor, 2006). This is especially the case with Aboriginal movements.

For the most part, the voices of Aboriginal peoples have been ignored, dismissed, or overridden since contact by the various levels of government within the country now called Canada. Similar stories to that of the Eeyouch (Cree) of James Bay have played out in Canadian history. In fact, much the same story is being played out now in northern Manitoba as hydroelectric development is being proposed, constructed, and opposed. Activist Sophia Rabliauskas of Poplar River First Nation (Manitoba) won international recognition and the 2007 Goldman Environmental Prize for North America for her work, to almost no mainstream media attention in Canada. (Former Grand Chief Coon Come won the same prize in 1994.) As Deacon and Baxter point out in their chapter, institutions such as the media have helped to make environmental injustice in Canada and those who oppose it less visible.

With respect to the Arctic and Subarctic, the Inuit have begun using the term "climate justice" to label climate change as a human rights issue (Watt-Cloutier, 2004). Chemical pollutants produced in industrial and energy production far to the south adversely affect northern food chains, as well as human and animal subsistence, and well-being. Northern peoples bear a disproportionate burden from economic development and natural resource extraction, since they do not benefit from the conveniences of mass industrial production as their southern neighbours do, yet experience deleterious changes to their economies, environments, and ways of life.

In this Introduction, we explore the context of environmental justice in at least two senses: as a conversation for community understanding, advocacy, and mobilization; and as a political principle that no self-identified group should bear a disproportionate burden of injustice (Agyeman and Evans, 2004). Canada is unique in that while it shares a continent with the United States and Mexico, it differs in many ways from these countries, including how environmental justice is imagined, manifested, (de)constructed, and shapeshifted from a predominantly Eurocentric/Eurogenic discourse (O'Riley and Cole, this volume) to alter-native ones.

Speaking for Ourselves is one coming-together of scholars on the relationship between environment and justice. We highlight the conversations relating to environmental justice espoused by the contributors rather than attempting to denote all that environmental justice can be, because we do not wish to define the parameters of the environmental justice discussion in Canada. No collection can be *the* definitive statement on the relationships between land, resources, nature, history, power relations, society, peoples, inequity, and all our (human and other-than-human) relations. Nevertheless, we hope that the stories shared will be part of a larger conversation in which the relations between humans, nonhumans, the environment, and social justice on the ground (and in the air and water) can be moved in a good way toward more equitable relations. In particular, *Speaking for Ourselves* takes its shape from the experiences and broad themes we have tried to incorporate as the collection developed.

Our Orientation as Editors

Environmental justice in Canada – as a transdiscipline in academia or a framing strategy in social policy and protest movements – acknowledges the particular relationships of the peoples of Canada with the land and with one another. From time immemorial (time before time, time before memory), First Peoples have been the inhabitants and stewards of what is now called Canada, with varied ways of conceiving of the relations of the beings of the land. From early European contact/invasion/occupation to the establishment of the Government of Canada as a sovereign nation in 1867 founded in the European tradition, imperial and colonial philosophies, policies, and practices have decimated First Peoples and their lands. The production of academically specific knowledge ("research" and "scholarship") has been carried out and used by socially dominant newcomers to further their social positions and agendas, and to shape and direct the trajectory of relations between all people in the cultural and social ecology of Canadian society. Western knowledge forms are very often treated by their creators as constituting the one true story of how the world works and what counts as legitimate knowledge. Whose knowledge counts? Whose intellectual and cultural properties are protected, and whose are not? Unwittingly or deliberately, academic scholarship subverts the very justice sought by so-called liberal societies, especially if elite and powerful actors within the mainstream are inconvenienced or discounted by it. From some perspectives, epistemological dominance manifests as social pre-eminence, and the exchange value of particular cultural capital is related to social and economic dominance and domination (Bourdieu, 1990). From an Aboriginal perspective, there is no "equivalency of epistemologies" (Grillo, 1998) on the Canadian landscape; academic, religious, and governmental Eurocentrism has made itself the default position of what counts as being legitimate knowledge and practice, dismissing or marginalizing "other"ed epistemologies, bodies, stories, and practices. Awareness of this inequity has led to the way that we have written this Introduction, and some of the choices we have made of chapters for inclusion in this book and contributions not selected.

Within the space of this book, we are striving to ensure that environmental justice is enacted – including that authors "speak for ourselves" rather than for/instead of others. This is one of the Principles of Environmental

Justice as established by the First National People of Color Environmental Leadership Summit in 1991.[2] There are as many environmental stories as there are storytellers. But who has been telling those stories? Whose voices have been heard, whose have been silenced? Who has copyrighted and published whose intellectual property (data/research findings) under the aegis of freedom of speech, academic freedom, and academic research (as an objectifying praxis)? How have these factors shaped the issues of social injustices and relationships to our environment(s)? Similar questions come up over and over. We felt that these issues needed to be told by people who are a part of the peoples and lands featured in the narratives.

Because of the pre-eminence Western nations place on the value of the written word, the written text becomes the de facto account. "Other" is cited and thereby fixed within a grasp/gaze. Each epistemologically facilitated research narrative comes from or through the particular cultural and social location, language, lenses, filters, memories, experiences, values, and accumulated cultural capital of the researcher-writer, and it is embedded in its society – its institutions and socially acceptable ways of knowing and doing things. We seek to unsettle such embeddedness, just as after hundreds of years of being designated as being nothing more than hearsay or unsubstantiated anecdotal evidence, oral narratives are becoming recognized by legal systems around the world as court case after court case affirms their legitimacy and validity.

There are many other voices that could have been included in this book, but as an editorial group, we chose at the outset to bring forward certain stories told by voices often made invisible by privileged outsider-experts and specifically by Western academic ways of knowing. This was especially so with regard to the circumstances of Aboriginal peoples. To allow space in which people can speak for themselves encourages a mutual intercultural conversation (Grillo, 1998), a culturally reciprocal space of speaking/listening. Working in respectful ways requires more time. The academy has existed for 150 years or so within the Canadian context and has been based on an unbalanced Aboriginal/non-Aboriginal power differential, with the latter being treated by the former as a source of data. There have long been requests that non-Aboriginal peoples walk beside or behind – but not in front of – Aboriginal peoples. Co-creating a space of shared storying not only brings a different kind of relationality within academia, it also encourages more equitable, diverse, complex, and complicating narrative engagements (Cole, 2006; O'Riley, 2004).

Environmentalism may be just a new word to describe an old concept or practice – caring for and about the world we live in. The social movement known as environmentalism has been very much implicated in new forms of colonialism and oppression globally, including within Canada. The prescriptive solutions of an expert, activist, or researcher not from the community suffering from environmental injustice may reproduce or magnify existing inequities, thereby contributing to disabling practices and disempowerment. Appropriation of voice by outsiders contributes to community and individual disempowerment. Aboriginal peoples have experienced this in abundance, to the extent that all academic research is considered highly suspect and, more often than not, disrespectful by many Aboriginal community members. Other communities have also felt the weight of the privileged academic and his (usually) epistemic dominance.

So, who are we to collect these stories here? A disparate crew to be sure, but not so disparate – we are academics all, each with a different activist background. Pat O'Riley was born in Quebec and is of Haudenosaunee, French, and Irish heritage; Julian Agyeman is of British and Ghanaian heritage, working in the United States; Peter Cole is Stl'atl'imx (British Columbia), with Welsh-Scottish heritage; Randy Haluza-DeLay is an American who traces his ancestry through Plymouth Plantation a decade after the first Pilgrims arrived from England; he immigrated to Canada two decades ago. We all teach at universities, and come to environmental justice through lived experience as well as through fields of study such as sociology, education, environmental studies, geography, Aboriginal and northern studies, theatre, and poetry. We have all lived and worked internationally. Agyeman has been active in urban sustainability issues on two sides of the Atlantic. O'Riley and Cole are partners and have lived and taught at universities in Canada, Aotearoa-New Zealand, and the United States. They have been working on language and cultural regeneration, economic renewal, and self-determination with Stl'atl'imx communities. Haluza-DeLay has lived in ten states and three provinces, has been a wilderness guide, and has worked for church-based peace and justice projects. O'Riley was involved in building design, labour, and human rights long before she did any post-secondary education. Since the 1960s, Cole has worked in all areas of landscaping and as a journalist and builder-renovator. Some of us are parents; all of us have families.

These social and cultural locators may hint at the variety of influences but do not even begin to describe the orientations we bring to this work. They affect the topics that have been collected, the processes by which we have selected and (hopefully) nurtured the contributors and the ways that we see environment, justice, and the joining together of those concepts and practices into the focus and momentum of this book: environmental justice.

Conceptualizing and Contesting Environmental Justice

The Canadian Environmental Protection Act (CEPA) of 1999 (the most recent incarnation of CEPA) states that the protection of the environment is essential to every Canadian's well-being. "Well-being" encompasses the provision of and access to environmental benefits such as good air quality, safe water, access to parks, natural spaces, public-transit options, quality housing, and reclamation of unceded First Nations' traditional lands and territories.

The laws of the land, along with individual and collective ethical responsibilities, also demand that Canadians be protected from environmental risks such as poor air quality, toxic chemicals, hazardous waste sites, and other environmental "bads." Research from many national contexts has consistently shown that exposure to these risks and externalities is disproportionately borne by poor and racialized peoples and communities (e.g., Agyeman, Bullard, and Evans, 2003). This has come to be known as "environmental injustice." Environmental well-being means access to aesthetic and healthy environments rather than just the avoidance of deprivation or harm. Thus, environmental concerns are *relative* to the population rather than universal or definitive standards of well-being or deprivation and exposure to harm.

One standard narrative is that community outrage at government policies and practices and industrially produced environmental hazards has been organized and mobilized in the form of civil society groups and joined by academic researchers, leading to the rise of the environmental justice (EJ) movement worldwide. This standard story needs contesting, as many of these movements have existed long before their (re)conceptualization as the EJ movement. Taylor (2000) notes that the field of *environmental justice* developed as a way of expressing environment-related social inequities. The term focuses on a wider range of social groups and issues than that covered by *environmental racism* (ER), which primarily names the disproportional impacts on communities of colour and indigenous communities in the United States. Numerous Canadian scholars believe that there has been little environmental justice-related research in Canada (Gosine, 2003; Draper and Mitchell, 2001; Teelucksingh, 2007) and very few Canadian books have been published using the environmental justice concept as a central analytic focus (Fletcher, 2003; Van Wynsberghe, 2002). One journal special issue (Haluza-DeLay, 2007) and a magazine issue (Gosine, 2003) have also been produced. This does not mean that the essence of EJ or ER has not been a focus of great concern for a very long time in this country. For example, since European contact, Aboriginal peoples have been articulating environmental injustices in relation to loss of land, Aboriginal title, and devastation of their traditional territories and the life forms they support. However, Aboriginal voices have rarely been heard amid the din of academic environmental theorizing. In their chapters, Keil, Ollevier, and Tsang, and Rahder show how EJ organizing has also occurred without explicit labelling as "environmental justice."

EJ and ER research has shown a clear pattern of racial and class demographics and a corresponding distribution of environmental risks or exposures (discussed below), but such distributive inequity is only one aspect of conceptualizing environmental justice. In (Eurocentric) legal theory, *procedural* justice – unequal participation in democratic processes as articulated in this volume by Ali, Reed, Ominayak, and Thomas – is another major type of justice. Those with limited access to information, participatory opportunities, and/or the power to shape discourse(s) or decisions are less able to defend themselves and their communities from negative distributional environmental effects, and less prepared to advocate successfully for better environmental conditions. For example, Pushchak (2002) detailed the process by which the federal government repeatedly conducted environmental impact assessments about the low-level military flights in Labrador that have caused hardship for Innu residents until an assessment favourable to the government's predetermined desires was finally reached. In yet another case, Hanson (2007) reported on farmers struggling against the power of a corporation's interests and resources in a manner that resonates with the tropes and modalities of procedural justice. As Agyeman (2005) comments, "The purveyors of environmental bads, such as large multinationals, are favored in pluralistic decision-making processes because of their disproportionate influence, economic muscle, and knowledge" (p. 3).

These examples illustrate the importance of what is called "procedural justice" as an integral facet of environmental inequity. Procedural justice theories can help to clarify the practices of "democratic" participation, including how social hierarchies participate in the construction of environmental injustice. Procedural inequities are more difficult to explain and measure – and change through policy and practice – than substantive or material inequities, so much so that Draper and Mitchell (2001) decided intentionally *not* to address it in a brief review of environmental justice in Canada.

Mainstream social justice literature has increasingly pointed out a third dimension beyond distributive and procedural justice issues (Barry and Eckersley, 2005; Schlosberg, 2004). This is the issue of *recognition* – such questions as Who has standing? On what concerns? What are the legitimation processes by which social and symbolic meaning are attributed to these questions? What constitutes power, and what are the contexts with respect to those with power and those without? These questions are fundamental to the quest for justice and equity but challenge the very foundations of legal process and often legal theory, which, as we have noted, is founded in particular cultural conditions such as those of the British or other European traditions. Several contributors in this book base their quest for justice on different foundations.

Justice as a Western construction sometimes includes the recognition of perceived differences, including how these perceived differences came to be constituted and how they are sometimes constitutive components of the issues themselves. For example, the perception of the Alberta prairies will differ for farmers several generations on the land (yet who do not "own" the subsurface rights), corporate transnational coal-bed methane producers, the provincial government, and Aboriginal communities that have been required to migrate from their traditional homelands to some other place which is already another community's (often unacknowledged) homeland.

Environmental justice issues do not involve only documentable technical data and scientific assumptions about the world but also fairness of treatment and the participatory ability of all marginalized peoples to be able to make substantive qualitative changes to the impositions of the larger society, especially those that adversely affect their rights and freedoms. Justice is broader than the liberal understanding of distributive equity (Fraser, 1997; Young, 1990; McGregor, this volume). "If differences are constituted in part by social, cultural, economic, and political processes, any examination of justice needs to include discussions of the structures, practices, rules, norms, language, and symbols that mediate social relations" (Schlosberg, 2004, p. 99). "Justice" should no longer be limited to Western legal concepts (though the word and the practice have deep Western roots) because Canada, as well as being Aboriginal land, is also a land of people who increasingly do not come only from the North Atlantic regions.

Environmental matters and justice matters are to a large extent about who gets to ask the questions, who gets to be heard (and listened to), and who benefits from how and if the questions are answered, researched, or considered relevant. For example, public transportation can be promoted as green – in terms of it being efficient and an environmental good – or as an equity issue – servicing people who require transportation but do not have automobiles. Is lack of public transit an environmental problem? Is it social? Is it injustice? How many levels and definitions of *(un)sustainable* are there and for whom are they (un)sustainable? Is a bus or subway sustainable for a bicyclist or committed pedestrian commuter? How big an ecological footprint is too big? Are economic growth and environmental sustainability mutually exclusive or reclusive? How are these concepts constituted and by whom?

Such questions are vital ones to ask. Although distributional inequality is an important conversation, exclusive focus on it narrows the scope and dynamic of justice. Similarly, inclusion of procedural justice dimensions – while an improvement – still limits participation to those who are allowed by the system itself to participate. It does not ask the more basic questions of who is the system, and how did the system come to be this way? Applying modern liberal rights discourse to the environment as an entity other than resources is difficult. The classic question, "do trees have legal standing?" challenges the common sense of modern anthropocentric "democracies" and modern systems that name nature in such a way as to reify it and make it into a commodity (a resource) for the benefit of the corporate entity (which has been given the legal status of "person"). In many Aboriginal cultures, rocks are persons, as are salmon and all of our relations. To whom does the oxygen produced by our photosynthesizing sisters and brothers belong and where is that negotiation taking place? In this regard, Anishnaabeg scholar and lawyer John Borrows (2002) asserts that "procedural realignments, through institutional reform and interest group reconfiguration," (p. 46) require not only the environmental knowledge of Aboriginal peoples to be taken into account, but also full participation of Aboriginal peoples, so that is it is less a (Western) legal discourse and more a conversation of interrelationship and interdependency of spirit, law, and cultural conventions. For true environmental justice, when will the West catch up?

Environmental discourse by those who care for and about the land has sought to extend recognition of value beyond narrow prescriptions/descriptions/inscriptions. It is no accident that Aboriginal peoples themselves have had to similarly battle for their survival and need to continually challenge the ongoing colonial system that has removed them from their traditional lands, as well as destroyed it through resource extraction, damming, and settler encroachment, to name a few injustices. As in most industrialized countries, history as taught in schools as a history of Canada ignores the tens of thousands of years Aboriginal peoples have inhabited this country (Dickason, 2006) and elaborates that history through white, Western, and Eurocentric eyes. Deliberately set-out state-enforced assimilationist policies and practices, unilateral erasure of treaty guarantees, denial of legal representation, denial or criminalization of spiritual practices, and other legalized and socialized forms of systemic violence amount to government-sanctioned (and often initiated) cultural and biological genocide. Aboriginal peoples have been rewriting these histories (e.g., Bird, 2007; Cole, 2006; Cole and O'Riley, 2007; Dickason, 2006; Lawrence, 2002; McMillan and Yellowhorn, 2004), thereby contributing to a conversation that has until recently been a monochrome monologue with an occasional dash of multicultural phrase or clause for colour.

Although "the environment" may seem, *prima facie,* obvious and self-explanatory, it is not so simple. What constitutes the "environment" in the contributions of this book varies considerably, as well as what constitutes "justice." Some EJ scholars feel that the notion of justice is undertheorized and, as a result, is weak in application. It must be noted that these conversations about land and justice have in large part been written using Eurocentric ways of thinking. Only in the past few decades have Canadians of non-European heritage been acknowledged in the environmental justice conversation (Cole, 1998; Gosine, 2003; Jafri, this volume; McGregor, this volume). Nor have economically disadvantaged people usually had the resources or contacts to write extensively on environment and justice – except, perhaps, in letters to the editor or as informants (rather than co-researchers) in academic research projects. The discussion already presented in this Introduction should help readers ponder the different ways that contributors are "languaging the land" (Cole, 2002). Finally, how contributors unite these concepts into claims about environmental justice is equally varied. We consider these variations an important part of the ongoing conversation on environmental justice in Canada.

Overview of Research on Environmental Justice in Canada

Most research in Canada is carried out by universities, governments, and transnational corporations (including media). The intellectual and cultural foundation of the bulk of publicly funded research in Canada needs to be questioned. How does it reinscribe culturally hegemonic (Eurocentric) standards of validity and legitimacy? What is the relationship between universities, academic journals, major funding councils, and transnational corporations, and what might this do to fundamental questions of economic-environmental justice? Which demographic groups have shaped the academy and retain power within it and the academic industry? Where do governments fit into this equation? Where do equity-seeking groups fit into the research processes? How is transparency and accountability ensured (or ignored)? Such questions are fundamental to the quest for social and environmental justice.

Canada has a unique ongoing colonial and racial history and social institutions. The patterns and relationships of academic and civil society discourses and dynamics are also unique. Aboriginal peoples, the most racially marginalized peoples in Canada, are most often left out of these "post"-colonial/anti-racist conversations. The very existence of the Aboriginal peoples of Canada is a challenge to the legitimacy of the Canadian state and its imported and imposed laws. Power determines legitimacy in law, especially when the law is itself derived almost wholly from European laws and values.

Canadian governmental policies impact patterns of exposure that particular groups have to environmental hazards, as well as just consideration within the justice system. The "social safety net" for the people of Canada should (in theory) mitigate some effects of environmental exposure; however, it has long been in the process of being dismantled (Carroll and Ratner, 2005) or replaced by a social safety net for transnational corporations. An official commitment (in practice) by the federal government to multiculturalism may affect the concentration of environmental risks, but there is little historical evidence for this because of the gulf between policy and practice. On the other hand, many Aboriginal peoples and other racialized groups in Canada see multiculturalism as being problematic – often little more than a pretense or diversion based on nebulous conceptions of equality (Das Gupta, 1999; Isin and Siemiatycki, 2002; Phillip, 1992).

The federal system of governance, with its delineation of responsibilities between provincial and federal jurisdiction, affects organizing for environmental well-being and social justice. The environment (natural resources, land management, endangered species, water quality) are provincial responsibilities. Health care is mandated as a national responsibility although delivered provincially. This division can affect the effectiveness of addressing environmental health issues. There are also many areas relating to human health and wellness that are lost in a limbo of inter-jurisdiction or layered jurisdiction, which cause a stalemate in terms of action on pressing environmental health issues. According to Rahder, Jafri, McGregor, Lawrence, Reed, and Cole and O'Riley in this book, traditional de/re/constructions of what counts as "environment" affects opportunities for organizing an environmental justice movement or addressing how and if environmental inequality occurs.

Global research has shown varied relationships between ethnic, racial, and class demographics and environmental inequality. Such research should be handled carefully in terms of its direct applicability to Canada or any other country. For example, Canada has different racial discourses and dynamics than the United States, our continental neighbour. Canada does not show such strong patterns of racial segregation, either regionally or in urban areas. Such urban dynamics may be changing, however, leading to the racializing of spaces and the concomitant potential for increased spatial inequality (Razack, 2002; Walks and Bourne, 2006).

The specific history of resource development and its place in regional and national economies also makes for specifically Canadian forms of environmental justice issues (Ali, this volume). Still, most prevalent of the environmental inequities within Canada may be the political and cultural position of First Nations. Aboriginal peoples are faced with systemic environmental injustice in terms of treaty and land claims processes; the situating of energy projects on or near their traditional territories; air, water, and land pollution; deplorable drinking-water quality issues (Indian and Northern Affairs Canada, 2003), resource extraction by outsiders on unceded territories by government-sanctioned contracts (Lovelace, this volume); the lack of ready and affordable access to economic development where they live; poor quality of life conditions, including access to education and health care; the failure by the Canadian state to recognize underlying and unalienable Aboriginal title and rights; and the unwillingness of the Canadian state to right historical wrongs to First Peoples.

Aboriginal peoples make up approximately 5 percent of the Canadian population. On nearly every marker of social inclusion or equity, they fare far worse than all other Canadians, with higher poverty rates, poorer housing, lower levels of formal education, lower employment levels, lower retention and promotion in employment, higher suicide and incarceration rates, and poorer health. There are, however, some indications that the situation is improving for some Aboriginal people (Ponting and Voyageur, 2001), at least in part because of increased political mobilization of Aboriginal groups and repeatedly winning legal and court recognition that treaty arrangements and unceded land holdings are still valid in contemporary Canada. The Royal Commission on Aboriginal Peoples (RCAP) (1995) was yet another study documenting the inequities, injustices, ongoing oppression, and structural violence experienced by Aboriginal peoples across Canada and throughout Canada's history. Disappointingly, more than ten years after the commission's report, there has been "a clear lack of action on the key foundational recommendations of RCAP and a resultant lack of progress on key socio-economic indicators" (Assembly of First Nations, 2006, p. 2). Such social processes affect environmental matters, with consequences that are key to contemporary environmental justice contestations and continuing social conditions. If a group is treated systemically and historically in such a way as amounts to the enactment of genocide against them, a foundation has been laid for dysfunctionalism within that group owing to the damage done over successive generations.

Finally, there is the environmental injustice that Canada, a wealthy nation (overall, but not for all) in the global North, wreaks on poorer countries and peoples in the global South. By having one of the world's largest per capita ecological footprints, Canadians are incurring an enormous ecological debt. Canada's global economic and political position indicates historical, contemporary, and future environmental injustice to the planet, its peoples, and all their/our relations. Social institutions such as transnational corporations, and so-called liberalization of trade policies, coupled with media reflecting corporate interests and promoting consumerist culture, combine to addle the citizenry. Some Canadians may be addicted to automobiles, big TVs, cheap fast-food from afar, and available energy, and some of us may allow environmental exploitation to occur because of those addictions, but the responsibility and the power to act differently are not equally distributed; therefore, the assumption of consumer sovereignty should be critiqued if it lacks a more comprehensive social analysis (Hanson, this volume).

Existing Canadian Environmental Justice Research

The existing research on Canadian environmental justice takes several forms. Most of it relies on government, industry, and university research data, but an increasing amount is done by NGO and ENGOs (environmental

NGOs) as well.[3] The first form of research is that of the aforementioned comparison of demographic markers (race, ethnicity, income, immigrant status, housing values, etc.) with proximity to known environmental exposure sites (identified toxics, polluting facilities, etc.). Generally, this line of research compares census data with databases such as the National Pollution Release Inventory for the same geographic unit, or tries to clarify potential relationships between health indicators such as asthma or cancer with environmental exposure locations. Such research is useful because it illuminates presence, prevalence, or absence of patterns of inequity and could communicate the breadth of potential problems in ways that may mobilize political support for addressing them. Such research, however, should not displace attention to impacts on individuals and communities in favour of the epistemic screen sometimes cast by the effort to have statistical science as the final arbiter. Furthermore, such exposures are typically multifactorial, which makes evidence gathering a methodologically difficult task (and this is not even problematizing the lack of information on what methods and methodologies were used in the gathering of statistical and polled data and how such data might be categorized and filtered). Such "public science" may have "unwittingly conservative effects" because it reproduces existing reliance on certain types of methodologies and denies others and establishes highly limiting understanding of satisfactory evidence and proof (Stacey, 2004, p. 131). In many ways, the methodological debates in social science correspond to recent legal debates in which oral history has been accorded legal standing, or the processes by which traditional ecological knowledge has reached some level of acceptability by bodies such as Environment Canada.

Other methodological problems abound in this demographic-proximity style of research. For example, census tracts do not match wind patterns, meaning that populations downstream or downwind from industrial releases and thus at greatest risk are not specifically captured by any currently collected data. Moreover, distribution correlations do not show the processes by which such disproportionality may have occurred, nor do they address fundamental questions of rights or power differentials. Much research of this nature has been done in the Hamilton area (Buzzelli, Jerrett, Burnett, and Finklestein, 2003; Buzzelli and Jerrett, 2004; Jerrett, Eyles, Cole, and Reader, 1997; Jerrett et al., 2001) and more recently Vancouver (Veenstra and Kelly, 2007), although Nabalamba (2005) also conducted a comparative study of three cities in the Golden Horseshoe area of Ontario. Finally, the environmental effects are usually those associated with human health, as Eyles discusses in his chapter, rather than a more comprehensive understanding of socio-environmental relationships.

These studies are highly limited in the number of studies completed, regions studied, and types of environmental exposure. Most importantly, these studies do not demonstrate causation, nor the social processes at work. Correlation of demographics with exposure could mean, for example, that the neighbourhood was already marginal, which then led to the siting of the offending facility (for the perceived economic benefit, or because of inability to mount a Not in My Backyard [NIMBY] campaign). Alternatively, the siting could have caused the lowering of property values. Despite these problems, we do see systemic patterns of environmental inequity, structured along dominant/dominating social patterns. Some socio-economic factors and environmental exposure seem related. These include poor housing (measured by housing value and state of repair relative to rental cost) and income. However, racial identification and immigrant status are contradictory, with some studies finding correlation and others finding none.

One response to these findings is that more studies of potentially patterned environmental risk with wider regional distribution are needed. However, another argument is that, although broad patterns discernible by quantitative measurement using databases not designed for such data extraction do not present definitive results, numerous case studies do illustrate environmental inequality in and between Canadian communities. Since environmental effects are primarily experienced in localities, case studies are useful to uncover these effects, including a grounded understanding of local places, their regional or global context, and the relations that link them. These relations can take the form of global air currents transporting pollutants from far-off production facilities, ideas about consumerism or conservation, governmental policies that allow environmental degradation to occur in local sites, or transnational economic systems that foreground profit over responsibility and stewardship. Inequalities also occur across localities, such as the 600-plus First Nations reserves, many of which do not have safe drinking water (Indian and Northern Affairs Canada, 2003). As they accumulate, the case studies are showing several foci.

First, much of the Canadian research has been *about* Aboriginal peoples (Cole, 2004; O'Riley, 2004). A pattern persists, from the mercury poisoning of the residents of Grassy Narrows and Whitedog reserves in northern Ontario (Erikson, 1995; Shkilnyk, 1985) to the toxic blob that affects Walpole Island in southern Ontario (Van Wynsberghe, 2002) to the preponderance of First Nations communities that have water-quality problems (Indian and Northern Affairs Canada, 2003). Mascarenhas (2007) assessed the intolerably poor water systems on Canadian reserves as a facet of neo-liberal colonization within the trajectory of historical oppression as government agencies have allowed their environmental duties to become privatized or otherwise off-loaded. Numerous researchers have pointed out how resource management (a Western take on environmental relations) ignores Aboriginal people's traditional fishing or land-use practices and cultural knowledge such that Western and scientific management practices are command-and-control impositions of power, leaving Aboriginal peoples still fighting regulation and marginalization (Braun, 2002; Gerwing and McDaniel, 2006; Page, 2007). Furthermore, Aboriginal lands, including forests, were never ceded to the Crown in British Columbia. Braun (2002) shows the discursive representations of the forest by environmentalists and the timber industry also exclude Aboriginal peoples, writing, "Struggles over the forests' cultural meaning are central to producing socially just post-colonial natures on Canada's west coast" (p. 25). From an Aboriginal perspective, there is no "post" in post-colonial. As Ominayak and Thomas and Lawrence write in their chapters, the marginalization of First Nations is an effect of state and corporate (including media and educational) strategies for appropriating land and exploiting or diverting natural resources from land use and occupancy to extraction and sale.

The pattern continues in the North, where chemical pollutants produced in industrial and energy production far to the south magnify and affect northern food chains, subsistence, and health and wellness (Trainor, Chapin III, Huntington, Kofinas, and Natcher, 2007; Trainor et al., this volume). Northern peoples bear a disproportionate burden, especially since they do not benefit directly from the industrial production yet experience the changes to their mixed economies. Specifically, the Inuit have organized against climate change in a campaign focusing on climate justice (Watt-Cloutier, 2004). Trainor et al. (2007) note that the policy regime in Canada allows the Inuit in Canada a better political position to shape federal policy than the Inuit in Alaska. "Better" is relative, however, and most would agree that Northerners wield little to no clout in the halls of Canadian power or over the implementation of policies decided in Ottawa, on Bay Street, or elsewhere nationally or internationally.

A second theme in Canadian environmental justice research is the *recognition* of Aboriginal difference, that is, that First Nations and other indigenous peoples have practices and epistemologies and relations with the land that bear little relationship to that of contemporary, *Westernized* Canadian society. This is especially prevalent in scholarship done by Aboriginal scholars and is among the reasons that this book takes the stand that it does regarding "speaking for ourselves." Furthermore, there is, and has long been, an ambiguous relationship between environmentalists and Aboriginal peoples. While environmentalists often appropriate Aboriginality as an exemplar of environmental praxis, this stereotypes Aboriginal peoples as well as essentializes them (Borrows, 2002). In this way, environmentalist discourse has a race-making element. Furthermore, environmentalist support does not necessarily extend to campaigns for land claims, treaty rights, or economic development, thus undermining cultural and social justice for

Aboriginal peoples (Ballamingie, 2006; Lawrence, this volume; Robinson, Tindall, Seldat, and Pechlaner, 2007; Stoddart, 2007).

A third theme in Canadian environmental justice research is the connection of race with the physical and social landscape (Jafri and Okamoto, 2007; Peake and Ray, 2001; Teelucksingh, 2002). As McCurdy (1995) reports it, the case of the dissolution of the black community of Africville in Nova Scotia is an example of racialized environmental injustice (see also Nelson, 2008). The North, the Canadian Shield, and "wilderness" play a significant narrative role in the Canadian imaginary. Analyses of the racialized colonial and neo-colonial construction of the land as empty requires continuing scrutiny of the often-ignored "whiteness" of Canadian national narratives and resource development (Baldwin, 2004; Cole, 2006).

A fourth theme is associated with health and safety, often involving toxic contamination. Canadian case studies of toxic contamination have often focused on industrial sites rather than on human bodies as repositories of industrial toxins. Reviewing six cases on toxic exposures, one set of researchers concluded that regardless of racialized environmental inequity, "the factor of class is invariably present" (Warrnier, McSpurren, and Nabalamba, 2001). In the Canadian tradition, class factors are predominated by location in the structures of the political economy and are not always separate from race issues. The case of the Sydney Tar Ponds in Halifax demonstrates this feature clearly. Working-class neighbourhoods experienced toxic loads far in excess of acceptable levels, while government authorities and their bureaucracies stalled, denied, asked for more research, and generally abrogated regulatory responsibility in favour of capital accumulation (Barlow and May, 2001; Thompson, 2003). Ali (2002, this volume) locates contamination in the workings of risk society and political economics, as do Haalboom, Elliott, Eyles, and Muggah (2006). Other cases involve uranium mining, most famously that of the Eldorado Mine at Port Radium on Great Bear Lake in the Northwest Territories. Documented in the video *A Village of Widows* (Blow, 1999), workers and their relatives in the Dene community of Deline claim that inadequate safety precautions have led to high cancer mortality, which is denied by government and corporate authorities.[4] In recent years, the effects of the extraction of oil from the Athabasca tar sands in northern Alberta has been claimed as a cause of health complications downstream. Aboriginal activists and the Prairie chapter of the Sierra Club have begun to use the term *environmental justice* to frame that issue, particularly as the provincial government has refused to conduct independent and peer-reviewed health studies. Although such cases often focus on health dimensions, there are other dimensions of social justice. Both Jackson and Curry (2004) and Harter (2004) argue that environmental struggles involve class interests, with asymmetrically situated procedural justice being part of the alternative to technocratic management of both people and resources. The social dislocation experienced by the swelling population in the tar sands-serving city of Fort McMurray seems to cry out for a political-ecological analysis that draws on the environmental justice and political economy literatures. These latter cases demonstrate how the intersection of environment and social considerations are part of larger political-economic and ecologic processes, especially in the increasingly global economy.

A fifth environmental justice research theme attempts to show the interplay between local environmental social issues and economic processes at larger scales. Debbané and Keil (2004) compare the provision of water services in South Africa and Canada. They highlight the rolling back of state services, specifically in moves to privatize water provision, which results in privatizing the water itself so that those less able to pay for services have less "right" to the water. They point out that environmental justice is not just abrogation of individual rights to environmental well-being but includes the ways that the state decides socio-environmental participation should occur.

Canadian scholars have a long political economy tradition, so it is little surprise that Canadian environmental justice scholarship includes a macroscopic analysis/restorying of societal restructuring in advanced capitalism. Such restructuring has included the retrenchment of governmental involvement in service provision, such as environmental protection (Paehlke, 2000). Another facet of this shrinkage includes the contracting out of services, and the off-loading of services onto other sectors of society, while increasing bureaucratic entanglements. Draper and Mitchell (2001) ask if the deficit in water quality that led to seven deaths and thousands of illnesses in Walkerton, Ontario, in 2000 is a case of environmental injustice. In Walkerton, the rights of citizens collided with the corporate agenda of profitability, with reduced governmental services even further reduced when handed to the private sector (Prudham, 2004). These are not random or accidental occurrences. Environmental regulation and public participation in development that impacts the environment have diminished in recent years (Carroll and Ratner, 2005; VanNijnattan, 1999). Environmental effects have increased as regulation and monitoring have been replaced by deregulation and "voluntary compliance," illuminating the state's compromised camaraderie with capital and "development."

Carroll and Ratner (2005) and their collaborators outline this capital-state partnership by drawing on numerous provincial case studies of social democracy in neoliberal times. The analyses specifically include environmental effects in the cases of Alberta, Saskatchewan, British Columbia, and Ontario, reinforcing the possibility that the modern state is interested only in providing environmental protection insofar as this allows the treadmill of production to proceed without catastrophic ecological collapse (Foster, 2002). Carroll and Ratner show how this has proceeded in Canada under assumptions of global competitiveness and a fatalistic sense of "we had to do it" in order to maintain sustained economic growth. Sustained, but not sustainable. The result is capital accumulation by those who own the means of production, rather than dispersion through the population of the presumed benefits of this economic activity. Governments have targeted ministries of environment for cutbacks – for all intents and purposes eliminating their ability to monitor corporate practices, much less provide proactive, distributive strategies for environmental improvements.

With government as capital's partner (willingly or "reluctantly," as Carroll and Ratner [2005] summarize), civil society groups, as well as racialized and Aboriginal peoples and communities, are left in the precarious situation of opposing powerful actors with even less means to do so. Community development is taking place in an increasingly inhospitable environment (Wharf, 1999, p. 267), which has only deteriorated at an accelerated rate. Procedural justice issues, then, are not just in the case of less educated or resourced citizens but are a facet of declining public access to political levers. The discussion of procedural justice scales up into the larger questions of the democratic deficit. Given this relationship between government and capital, and government's support of so-called economic development while conceding to its citizens a level of environmental protection only as necessary for the maintenance of the economic growth engine, it is increasingly unlikely for governments to adequately defend against environmental degradation (Pellow and Brulle, 2005). Nevertheless, many believe that the state still has more ability than any other actor or force to regulate the actions of other actors and the treadmills of consumption and production (Barry and Eckersley, 2005). To address environmental degradation and how it disproportionately affects some social groups more than others, we need a sound assessment of the macroscopic forces in operation. Canada has long been a follower (at best) rather than a leader in terms of environmental responsibility and action, even though such inaction causes much hardship or even death for significant numbers of its population.

These themes, as well as others, are represented in this book. Many of the contributors do not use the explicit language of environmental justice. We have deliberately allowed that, as our intention is to demonstrate environmental justice in Canada as being a dynamic and organic conversation rather than one using a prescribed rhetoric; certainly, this conversation has been going on since European contact and no doubt before. This process also facilitates dialogue on the diversity and multiplicity of intersections of environmental matters and social justice that is specific to the Canadian context. In this manner and in many other ways, a more global, inclusive, reflexive, and polyvocal approach can be put into practice. As citizens of the world, Canadians are directly affected by environmental injustice of sisters and brothers elsewhere in the world.

Introducing the Chapters

This book begins with the voices of the Original Peoples of what is now called Canada. Robert Lovelace, retired chief of Ardoch Algonquin First Nation, Deborah McGregor (Anishnaabe), and Bonita Lawrence (Mi'kmaq) share what "environment" and "justice" mean from Aboriginal perspectives and how they are enacted, as well as elaborating on struggles by Aboriginal peoples to reclaim and protect their traditional lands from imperialist, colonialist, and capitalist exploitation. Bob Lovelace wrote his "Notes from Prison" while serving in a maximum-security prison his sentence for trying to peacefully protect Algonquin lands from uranium mining. Ironically, only when imprisoned did Bob have the time to write, scribing his piece with a four-inch-long pencil sharpened on the side of a metal staircase. Bob, along with several other potential contributors, could not write chapters in time for the deadline for this book because they were too busy on the front lines of environmental justice.

McGregor writes that "relationships based on environmental justice are not limited to relations between people but consist of those among all beings of creation." McGregor roots this perspective in natural law as understood within the Anishnaabe world view. This view is rooted in creation and re-creation, which requires both the Creator and an ongoing process of relationship between all parts of creation.

Lawrence details Mi'kmaq land relationships in Newfoundland. Through the use of interviews, she explains how land is central to identity for indigenous peoples – often in ways difficult for non-Aboriginal Canadians to understand. Until 2008, Newfoundland Mi'kmaq communities were not "recognized" by the federal government under the Indian Act. More trenchantly, Lawrence contests the popular and academic history that asserts that Mi'kmaq are not "native" to Newfoundland. This contested history is part of the environmental injustice.

Roger Keil, Melissa Ollevier, and Erica Tsang begin their chapter by contesting the typical story that there is little of a social movement for environmental justice in Canada. In particular, they address that narrative by narrowing their scope to a detailed analysis of the City of Toronto. They conclude that "there are – substantially, if not by name – many environmental justice initiatives underway in Toronto."

In her chapter on gender and the environment, Barbara Rahder makes a similar argument. Rahder considers that environmental justice issues are feminist concerns and that women do not self-identify exclusively with an "environmental" label. She argues that their involvement in intersected social-environmental activism is an application of more thorough-going citizenship that overcomes other forms of social exclusion.

The Sydney, Nova Scotia, tar ponds are the primary case study for S. Harris Ali. He develops a theoretical account of environmental injustice that attends to the particular manifestations of this case but places them in larger socio-historical and political-economic contexts. "The broader forces involved in the Canadian political economy expose those in particular regions to harm," Ali writes. As his case study shows, criticisms of policies and practices that give rise to environmental harm are often muted by claiming them as localized manifestations rather than part of a systemic problem.

Similar processes are occurring with the Lubicon Cree First Nation of central Alberta. The account, by Chief Bernard Ominayak and Kevin Thomas, generates key questions for conceptualizing the processes of Canadian environmental justice: When did the manifestation of environmental injustice begin? The years 1979, 1939, and 1899, and even centuries past, are all potential answers. The rapid expansion of oil and gas drilling on Lubicon traditional-use lands continues to produce dramatic degradation in Lubicon social sustainability, economic well-being, and health, along with large-scale changes of the land.

Environmental justice activism has often revolved around health concerns. Good health is not equally distributed across the population. John Eyles situates environmental justice within the domain of population health, and its practitioners' approach to improving health by focusing on the social determinants that reduce healthiness. Although public health agencies consider physical environment as among the important determinants of health for population groups, Eyles argues that they do not consider it enough. He refers to the methodological assumptions of Western science and concludes that methodological and conceptual innovation is needed.

In many ways, Sarah Fleisher Trainor and her collaborators pick up where Eyles leaves off. Focusing on the Arctic, their chapter also addresses health and environmental injustice. The presence of persistent organic pollutants (POPs) and warming from global climate change mean that "the longevity and integrity of cultural traditions are thus intimately linked to the integrity and health of a functioning ecosystem," the authors point out.

There are few easy answers to combining justice and sustainability, as Maureen G. Reed shows in her chapter comparing community-based management systems at two Canadian UNESCO Biosphere Reserves: Redberry Lake in Saskatchewan and Clayoquot Sound in British Columbia. Reed's narrative shows the importance of considering participatory injustice, power relations (even among ENGOs), and local context in order to improve justice and equity in environmental management.

The media shape and sometimes distort public understanding of the world, so any imagining of environmental justice might begin with an understanding of how environment and inequity are presented in public media. Leith Deacon and Jamie Baxter provide the results of their content analysis of the two main national newspapers from 1986 to 2006. Their data also provide insight into what sources of information are cited and valued, all of which are important to raising the public profile of intersections of justice and the environment.

Lorelei Hanson considers the food system as part of the agenda of environmental justice and scrutinizes the five biggest Canadian ENGOs to see how they address food issues. Her assessment shows they have some recognition of the social and environmental implications of the food-production system but that "the examination of the overall system . . . is largely missing." The result is an over-reliance on advocating for personal lifestyle choices and consumer sovereignty, which cannot address the corporatization of agriculture that affects both people and land.

Beenash Jafri, a seasoned anti-racist activist, articulates some of the race erasing that occurs in environmental organizations. By examining the discourses of three ENGOs and interviewing staff (all in Toronto), Jafri shows how multiculturalism and "diversity" are substituted for anti-racism and lack analysis of the structural conditions of ENGO member homogeneity, white privilege, or colonialism follows. Jafri concludes that crucial for the re-imagining of environmental justice is to look "beyond the issue of the exclusion of people of colour within the environmental movement . . . to the multifarious ways in which environmental organizing strategies might be supporting relations of domination and subordination."

The last words in this section, and the book, are those of tricksters, Coyote and Raven (a.k.a. Pat O'Riley and Peter Cole). This chapter enacts an Aboriginal storytelling of environment and justice, a quite different storytelling from Western discourses. Cole and O'Riley join the circle begun by Lovelace in the prologue of *Speaking for Ourselves: Environmental Justice in Canada*.

As the chapters in the book fold and unfold, join and diverge, readers are offered an expanded, many-faceted, and polyvocal appreciation of and effort for environmental justice in Canada. We invite other positions and advocates, even as we position ourselves with the assertion that advocacy and scholarship must both be performed to make a world that is just and sustainable, for all of human and more-than-human peoples. Woe be unto us who unthinkingly add field to field, house to house, and carbon atom and POP to atmosphere and water until we find we live alone in a despoiled land. Let the ravens cry out, the coyotes howl, and the trees clap their hands that the humans might come to their senses.

Notes

1 See the archives of the Canadian Broadcasting Corporation at http://archives.cbc.ca/IDD-1-69-94/life_society/james_bay/.
2 The list of seventeen principles is widely available, see http://www.ejnet.org/ej/principles.html.
3 NGO research, which may be funded from the public purse, should not automatically be considered either neutral or circumspect.
4 See also http://www.sombake-themoneyplace.com/.

References

Agyeman, J. 2005. *Sustainable Communities and the Challenge of Environmental Justice.* New York: New York University Press.

Agyeman, J., R.B. Bullard, and B. Evans (Eds.). 2003. *Just Sustainabilities: Development in an Unequal World.* London: Earthscan Publications/MIT Press.

Agyeman, J., and B. Evans. 2004. "Just Sustainability": The Emerging Discourse of Environmental Justic in Britain? *Geographical Journal, 170*(1), 155-64.

Ali, S.H. 2002. Disaster and the Political Economy of Recycling: Toxic Fire in an Industrial City. *Social Problems, 49*(2), 129-49.

Assembly of First Nations. 2006. *Royal Commission on Aboriginal Peoples at Ten Years: A Report Card.* Ottawa: Assembly of First Nations. http://afn.ca/cmslib/general/afn_rcap.pdf.

Baldwin, A. 2004. An Ethics of Connection: Social-Nature in Canada's Boreal Forest. *Ethics, Place and Environment, 7*(3), 185-94.

Ballamingie, P. 2006. Paradoxes of Power: The Lands for Life Public Consultations (Ph.D. dissertation, Carleton University, Ottawa).

Barlow, M., and E. May. 2001. *Frederick Street: Life and Death on Canada's Love Canal.* Toronto: HarperCollins.

Barry, J., and R. Eckersley (Eds.). 2005. *The State and the Global Ecological Crisis.* Cambridge, MA: MIT Press.

Bird, L. 2007. *The Spirit Lives in the Mind: Omushkego Stories, Lives and Dreams.* Montreal: McGill-Queen's University Press.

Blow, P. (Director). 1999. *The Village of Widows: The Story of the Sahtu Dene and the Atomic Bomb* [videorecording]. Peterborough, ON: Lindum Films.

Borrows, J. 2002. *Recovering Canada: The Resurgence of Indigenous Law.* Toronto: University of Toronto Press.

Bourdieu, P. 1990. *In Other Words: Essays toward a Reflexive Sociology.* Stanford, CA: Stanford University Press.

Braun, B. 2002. *The Intemperate Rainforest: Nature, Culture, and Power on Canada's West Coast.* Minneapolis: University of Minnesota Press.

Buzzelli, M., and M. Jerrett. 2004. Racial Gradients of Ambient Air Pollution Exposure in Hamilton, Canada. *Environment and Planning A, 36*(10), 1855-77.

Buzzelli, M., M. Jerrett, R. Burnett, and N. Finklestein. 2003. Spatiotemporal Perspectives on Air Pollution and Environmental Justice in Hamilton, Canada, 1985-1996. *Annals of the American Association of Geographers, 93*(3), 557-73.

Carroll, W.K., and R.S. Ratner. 2005. *Challenges and Perils: Social Democracy in Neoliberal Times.* Halifax: Fernwood.

Cole, P. 2006. *Coyote and Raven Go Canoeing: Coming Home to the Village.* Montreal: McGill-Queen's University Press.

–. 2004. Trick(ster)s of Aboriginal Research: Or How to Use Ethical Review Strategies to Perpetuate Cultural Genocide. *Native Studies Review, 15*(2), 7-30.

–. 2002. Land and Language: Translating Aboriginal Cultures. *Canadian Journal of Environmental Education, 7*(1), 67-84.

–. 1998. An Academic Take on Indigenous Traditions and Ecology. *Canadian Journal of Environmental Education, 3,* 100-15.

Cole, P., and P. O'Riley. 2007. Coyote and Raven Talk about the Land/scapes. In M. McKenzie, H. Bai, P. Hart, and B. Jickling (Eds.), *Fields of Green: Re-storytelling Education* (pp. 121-34). Cresskill, NJ: Hampton Press.

Das Gupta, T. 1999. The Politics of Multiculturalism: "Immigrant Women" and the Canadian State. In E. Dua and A. Robertson (Eds.), *Scratching the Surface: Canadian Anti-Racist, Feminist Thought* (pp. 187-205). Toronto: Women's Press.

Debbané, A.-M., and R. Keil. 2004. Multiple Disconnections: Environmental Justice and Urban Water in Canada and South Africa. *Space and Polity, 8*(2), 209-25.

Dickason, O.P. 2006. *A Concise History of Canada's First Nations.* Toronto: Oxford University Press.

Draper, D., and B. Mitchell. 2001. Environmental Justice Considerations in Canada. *Canadian Geographer, 45*(1), 93-98.

Erikson, K. 1995. *A New Species of Trouble: The Human Experience of Modern Disasters.* New York: Norton.

Fletcher, T. 2003. *From Love Canal to Environmental Justice: The Politics of Hazardous Waste on the Canada – US Border.* Peterborough, ON: Broadview Press.

Foster, J.B. 2002. *Ecology against Capitalism.* New York: Monthly Review Press.

Fraser, N. 1997. *Justice Interruptus: Critical Reflections on the Postsocialist Condition.* London: Routledge.

Gosine, A. 2003. Myths of Diversity. *Alternatives, 29,* 4-8.

Grand Council of the Crees (Eeyou Astchee). 1998. *Never Without Consent: James Bay Crees Stand against Forcible Inclusion into an Independent Quebec.* Toronto: ECW Press.

Grillo, E.F. 1998. Development or Cultural Affirmation in the Andes? In F. Apffel-Marglin (Ed.), *The Spirit of Regeneration: Andean Culture Confronting Western Notions of Development* (pp. 124-45). London: St. Martin's Press.

Haalboom, B., S.J. Elliott, J. Eyles, and H. Muggah. 2006. The Risk Society at Work in the Sydney "Tar Ponds." *Canadian Geographer, 50*(2), 227-41.

Haluza-Delay, R.B. (Ed.). 2007. "Environmental Justice in Canada." *Local Environment 12*(6), 557-64.

Hanson, L.L. 2007. Environmental Justice across the Rural Canadian Prairies: Agricultural Restructuring, Seed Production and the Farm Crisis. *Local Environment, 12*(6), 599-611.

Harter, J.-H. 2004. Environmental Justice for Whom? Class, New Social Movements, and the Environment: A Case Study of Greenpeace Canada, 1971-2000. *Labour/Le Travail 54,* 83-119.

Indian and Northern Affairs Canada. 2003. *National Assessment of Water and Wastewater Systems in First Nations Communities.* Ottawa: Government of Canada.

Isin, E.F., and M. Siemiatycki. 2002. Making Spaces for Mosques: Struggles for Urban Citizenship in Diasporic Toronto. In S. Razack (Ed.), *Race, Space and the Law: Unmapping a White Settler Society* (pp. 185-209). Toronto: Between the Lines.

Jackson, T., and J. Curry. 2004. Peace in the Woods: Sustainability and the Democratization of Land Use Planning and Resource Management on Crown Lands in British Columbia. *International Planning Studies, 9*(1), 27-42.

Jafri, B., and K. Okamoto. 2007. Green Is Not the Only Colour: Reflections on the State of Anti-Racist Environmentalism in Canada. *Briarpatch Magazine* (December/January). http://www.briarpatchmagazine.com/news/?p=514.

Jerrett, M., R. Burnett, P. Kanaroglou, J. Eyles, N. Finkelstein, C. Giovis, and J. Brook. 2001. A GIS-Environmental Justice Analysis of Particulate Air Pollution in Hamilton, Canada. *Environment and Planning A, 33,* 955-73.

Jerrett, M., J. Eyles, D. Cole, and S. Reader. 1997. Environmental Equity in Canada: An Empirical Investigation into the Income Distribution of Pollution in Ontario. *Environment and Planning A, 29,* 1777-800.

Lawrence, B. 2002. Rewriting Histories of the Land. In S. Razack (Ed.), *Race, Space and the Law: Unmapping a White Settler Society* (pp. 21-46). Toronto: Between the Lines.

MacGregor, S. 2006. *Beyond Mothering Earth: Ecological Citizenship and the Politics of Care.* Vancouver: UBC Press.

Mascarenhas, M. 2007. Where the Waters Divide: First Nations, Tainted Water, and Environmental Justice in Canada. *Local Environment, 12*(6), 565-77.
McCurdy, H. 1995. Africville: Environmental Racism. In L. Westra and P. Wenz (Eds.), *Faces of Environmental Racism: Confronting Issues of Global Justice* (pp. 75-92). Lanham, MD: Rowan and Littlefield.
McMillan, A., and E. Yellowhorn. 2004. *First Peoples in Canada.* Vancouver: Douglas and McIntyre.
Nabalamba, A. 2005. Locating Risk: A Multivariate Analysis of the Spatial and Socio-Demographic Characteristics of Pollution. Paper presented at the annual meeting of the Canadian Sociology and Anthropology Association, London, ON.
Nelson, J. 2008. *Razing Africville: A Geography of Racism.* Toronto: University of Toronto Press.
O'Riley, P. 2004. Research and Aboriginal Peoples: Toward Self-Determination. *Native Studies Review, 15*(2), 7-96.
Paehlke, R. 2000. Environmentalism in One Country: Canadian Environmental Policy in an Era of Globalization. *Policy Studies Journal, 28*(1), 160-76.
Page, J. 2007. Salmon Farming in First Nations' Territories: A Case of Environmental Injustice on Canada's West Coast. *Local Environment, 12*(6), 613-26.
Peake, L., and B. Ray. 2001. Racializing the Canadian Landscape. *Canadian Geographer, 45*(1), 93-98.
Pellow, D.R., and R. Brulle. 2005. *Power, Justice, and the Environment: A Critical Appraisal of the Environmental Justice Movement.* Cambridge, MA: MIT Press.
Phillip, M.N. 1992. Why Multiculturalism Can't End Racism. *Frontiers: Selected Essays and Writings on Racism and Culture, 1984-1992* (pp. 181-86). Toronto: Mercury Press.
Ponting, R., and C. Voyageur. 2001. Challenging the Deficit Paradigm: Grounds for Optimism among First Nations in Canada. *Canadian Journal of Native Studies, 21*(2), 275-307.
Prudham, S. 2004. Poisoning the Well: Neoliberalism and the Contamination of Municipal Water in Walkerton, Ontario. *Geoforum, 35,* 343-59.
Pushchak, R. 2002. Environmental Justice and the EIS: Low-level Military Flights in Canada. *International Journal of Public Administration, 25*(2/3), 169-91.
Razack, S. (Ed.). 2002. *Race, Space and the Law: Unmapping a White Settler Society.* Toronto: Between the Lines.
Robinson, J.L., D.B. Tindall, E. Seldat, and G. Pechlaner. 2007. Support for First Nations' Land Claims amongst Members of the Wilderness Preservation Movement: The Potential for an Environmental Justice Movement in British Columbia. *Local Environment 12*(6), 579-98.
Royal Commission on Aboriginal Peoples. 1995. Ottawa: Government of Canada.
Schlosberg, D. 2004. Reconceiving Environmental Justice: Global Movements and Political Theories. *Environmental Politics, 13*(3), 517-40.
Shkilnyk, A.M. 1985. *A Poison Stronger Than Love: The Destruction of an Ojibway Community.* New Haven, CT: Yale University Press.
Stacey, J. 2004. Marital Suitors, Court Social Science Spin-sters: The Unwittingly Conservative Effects of Public Sociology. *Social Problems, 51*(1), 131-45.
Stoddart, M.C.J. 2007. "British Columbia Is Open for Business": Environmental Justice and Working Forest News in the *Vancouver Sun. Local Environment 12*(6), 663-74.
Taylor, D. 2000. The Rise of the Environmental Justice Paradigm: Injustice Framing and the Social Construction of Environmental Discourses. *American Behavioral Scientist, 43*(4), 508-80.
Teelucksingh, C. 2002. Spatiality and Environmental Justice in Parkdale (Toronto). *Ethnologies, 24*(1), 119-41.
–. 2007. Environmental Racialization: Linking Racialization to the Environment in Canada. *Local Environment, 12*(6), 645-61.
Thompson, S. 2003. From a Toxic Economy to Sustainability: Women Activists Taking Care of Environmental Health in Halifax. *Canadian Women Studies, 23*(1), 108-14.

Trainor, S.F., F.S. Chapin III, H.P. Huntington, G. Kofinas, and D.C. Natcher. 2007. Arctic Climate Impacts and Cross-Scale Linkages: Environmental Justice in Canada and the United States. *Local Environment, 14*(6).

VanNijnattan, D.L. 1999. Participation and Environmental Policy in Canada and the United States: Trends over Time. *Policy Studies Journal, 27*(2), 267-87.

Van Wynsberghe, R.M. 2002. *Alternatives: Community, Identity and Environmental Justice on Walpole Island.* Toronto: Allyn and Bacon.

Veenstra, G., and S. Kelly. 2007. Comparing Objective and Subjective Status: Gender and Space (and Environmental Justice?). *Health and Place, 13*(1), 57-71.

Walks, R.A., and L.S. Bourne. 2006. Ghettos in Canada's Cities? Racial Segregation, Ethnic Enclaves and Poverty Concentrations in Canadian Urban Areas. *Canadian Geographer, 50*(3), 273-97.

Warrnier, G.K., K. McSpurren, and A. Nabalamba. 2001. Social Justice and Environmental Equity: Distributing Environmental Quality. *Environments, 29*(1), 85-99.

Watt-Cloutier, S. 2004. Climate Change and Human Rights. In *Human Rights Dialogue: Environmental Rights Series, 2*(11). http://www.carnegiecouncil.org/resources/publications/dialogue/2_11/section_1/4445.html.

Wharf, B. 1999. Editorial Introduction: Community Development in Canada. *Community Development Journal, 34*(4), 267-69.

Young, I.M. 1990. *Justice and the Politics of Difference.* London: Routledge.

1
Honouring Our Relations: An Anishnaabe Perspective on Environmental Justice

Deborah McGregor

My goal in this chapter is to discuss environmental justice from a First Nations perspective, using water issues as a specific example. I wish to do so from my own standpoint as an Anishnaabe woman from Wiigwaaskinga (Birch Island, Ontario).[1] As I learn more teachings about water and the concept of "all our relations" I have come to understand that relationships based on environmental justice are not limited to relations between people but consist of those among all beings of Creation. From the perspective of the world view within which I am embedded, environmental justice is most certainly about power relationships among people and between people and various institutions of colonization. It concerns issues of cultural dominance, of environmental destruction, and of inequity in terms of how certain groups of people are impacted differently by environmental destruction from others, sometimes by design. But environmental justice from an Aboriginal perspective is more than all of these. It is about justice for all beings of Creation, not only because threats to their existence threaten ours but because from an Aboriginal perspective justice among beings of Creation is life-affirming. Aboriginal authors such as Anishnaabe environmental activist Winona LaDuke refer to this as "natural law" (LaDuke, 1994). While people certainly have a responsibility for justice, so do other beings (e.g., water and medicinal plants).

This aspect of relationships has not been adequately explored in contemporary discourse on environmental justice, although it is discussed at length in traditional teachings. The environmental justice literature instead assumes a certain ideology about "environment" – one with a focus on how certain groups of people (especially those bearing labels such as "minority," "poor," "disadvantaged," or "Native") are impacted by environmental destruction, as if the environment were somehow separate from us. Current discourse, and policy direction now in place (in the United States, at least; Canada as yet has no environmental justice policy), ignore Creation's responsibilities and duties to ensure justice. In the Anishnaabe world view, all beings of Creation have spirit, with duties and responsibilities to each other to ensure the continuation of Creation. Environmental justice in this context is much broader than "impacts" on people. There are responsibilities beyond those of people that also must be fulfilled to ensure the processes of Creation will continue.

Environmental justice is frequently presented as a relatively new concept, both in North America and internationally. Aboriginal people, however, hold ancient and highly developed ideas of justice that have significant applicability in this area. This chapter will explore concepts of environmental justice from such an Indigenous knowledge perspective.

The main theme of this chapter involves explaining the concept of environmental justice from an Anishnaabe point of view. An Anishnaabe understanding of environmental justice considers relationships not only among people but also among *all our relations* (including all living things and our ancestors). Environmental *in*-justice, then, is not only inflicted by dominant society upon Aboriginal peoples, people of colour, and people in low-income neighbourhoods but also upon Creation itself. To say that there is no recognition in the environmental justice discourse of this understanding would be incorrect. For example, the Principles of Environmental Justice developed by delegates at the First National People of Color Environmental Leadership Summit in October 1991 proclaim the right of all species, not just people, to be free from ecological destruction. This highlights the fact that both people and other species are impacted by environmental destruction. These principles also recognize the unique situation of Native peoples in the United States in terms of their legal relationship to the US government (Washington Office of Environmental Justice, 1991). Despite such proclamations, the topic of environmental justice with respect to Creation is not adequately addressed. Exploring an Anishnaabe perspective of environmental justice offers a unique opportunity to broaden discussion and understanding of this subject.

To adequately consider environmental justice from an Anishnaabe or other Indigenous knowledge framework, we must go beyond conventional discourse on environmental justice. Some of the best examples of work that begins to do this include Jace Weaver's *Defending Mother Earth: Native American Perspectives on Environmental Justice*. Based on a conference on the same topic, this work illustrates unique aspects of Indigenous peoples' struggles in relation to assaults on Mother Earth. Weaver (1996) notes that, in contrast to mainstream discourse, "discussion of environmental justice from a Native perspective requires an analysis of sovereignty and the legal framework that governs environmental matters in Indian country" (p. 108). For Native people, environmental justice is about colonization and racism, and not only about continued assaults on the environment (p. 107). It is recognized that Aboriginal people have responsibilities to fulfil that will ensure harmonious relationships with Creation. As Russell Means (1996) states, "We are the children of the Earth. She is our Mother, and it's our right and duty to protect her . . . From our traditional ways, we know that we do not have the right to degrade our Mother and that we must live in harmony with all of Creation" (p. xii).

Winona LaDuke has produced a variety of works in this area. *All Our Relations* (1999) celebrates Native environmentalism by chronicling various environmental injustice issues in Canada and the United States. *Recovering the Sacred* (2005), expands the scope of environmental justice in relation to Indigenous peoples by addressing such topics as biopiracy, exploitive research, theft of Aboriginal culture, and repatriation. Back in the 1990s, she wrote the Foreword, "A Society Based on Conquest Cannot Be Sustained," to Al Gedicks' book *The New Resource Wars: Native and Environmental Struggles against Multinational Corporations,* which is devoted to the topic of environmental racism. In it, LaDuke (1994) states that Native people "find themselves the target of industrialism's struggle to dominate the natural world; [Native people] are possessed of resources, lands and waters, now demanded by urban areas and industrial machinery often thousands of miles distant" (p. x). Other examples of environmental injustice issues relating to Aboriginal people in Canada involve hydroelectric development, mining, nuclear-waste siting, and low-level military flying. Aboriginal people continue to battle these issues, often with few allies. As Matthew Coon Come states, "We will be a voice for social and environmental justice for Aboriginal peoples, and for the animals and the land. We will be a voice for honour in dealings between governments and indigenous peoples. And we will be a voice against racist double standards that continue to oppress us, and continue to dishonour those in whose names they are used against us" (Coon Come, 1995, p. 16).

These issues are not always placed specifically in the environmental justice literature (though there are noteworthy examples that do, for example, Ashini, 1995; Pushchak, 2002).

Natural Law and Environmental Justice

Current environmental laws are inadequate for protecting what is important to Aboriginal peoples (Assembly of First Nations, 1993; RCAP, 1996). This is reflected in LaDuke's work (1994, p. x), where she argues that natural law, which has existed for thousands of years, is a source of justice that has served Aboriginal people for thousands of years and that can continue to do so. Anishnaabe legal scholar John Borrows (2002), in his book *Recovering Canada: The Resurgence of Indigenous Law,* states that "Aboriginal peoples developed spiritual, political and social conventions to guide their relationships with each other and with the natural environment. These customs and conventions became the foundations of many complex systems of government and law" (p. 47). Referring specifically to the Anishnaabe, Borrows indicates the seriousness of failing to abide by such natural laws: "If the Anishnabek do not honour and respect their promises, relations and environments, the eventual consequence is that these resources will disappear. When these resources are gone, no matter what they are, the people will no longer be able to sustain themselves because . . . while the resources have an existence without us, we have no existence without them" (p. 20). From an Aboriginal perspective, there is a clear need to reaffirm our understanding of natural laws to ensure the continued existence of all of Creation. Exploring these natural laws, where they come from, and the insights that can be gained from particularly the Anishnaabe tradition in this case is the focus of the remainder of this chapter. My chapter will draw upon a number of Anishnaabe thinkers who have influenced my work and helped frame my thinking around natural law, Indigenous knowledge, and environmental justice.

Environmental Justice and the Anishnaabe

Environmental justice is not a new concept in Anishnaabe culture. Natural laws have existed for generations that ensure justice for "all our relations." In contrast to perhaps more mainstream writings of environmental justice, all beings of Creation will be described in this chapter as having agency and entitlement, according to Anishnaabe tradition. As well, the ancestors of current beings and those yet to come (at least as far ahead as seven generations from now) also have entitlement to environmental justice. From an Anishnaabe perspective, the spirit world and all beings of Creation, including people, have relationships and responsibilities. Anishinaabe legal scholar Darlene Johnston (2006) states: "In Anishnaabeg culture, there is an ongoing relationship between the Dead and the Living; between Ancestors and Descendants" (p. 17). Anishnaabeg have to routinely consider questions such as, What is our relationship with our ancestors? Are we honouring our relationships with our ancestors? Are we doing justice to our ancestors and to those yet to come? The Anishnaabeg had codes of conduct and practices that ensured such relations would remain harmonious. To explore these questions, it is important to understand the Anishnaabe world view. This chapter will therefore begin at the beginning, with Anishnaabe Creation stories. There is more than one Anishnaabe point of view.

Creation

> Origin stories say a great deal about how people understand their place in the universe and their relationships to other living things. I have been taught by Anishnaabeg Elders that all Creation Stories are true.
>
> – D. Johnston,
> *Connecting People to Place*, p. 7

The Anishnaabe Creation and Re-Creation stories inform us of our beginnings and provide the conceptual frameworks for an Indigenous understanding of our relationship to Creation and its many beings. Anishnaabe storytellers Basil Johnston, author of *Ojibway Heritage,* and Edward Benton-Banai, author of *The Mishomis Book,* both begin their books with Creation. Here is part of the story as told by Johnston:

> Kitche Manitou (The Great Spirit) beheld a vision. In this vision he saw a vast sky filled with stars, sun, moon, and earth. He saw an earth made of mountains and valleys, islands and lakes, plains and forests. He saw trees and flowers, grasses and vegetables. He saw walking, flying, swimming, and crawling beings. He witnessed the birth, growth and the end of things. At the same time he saw other things live on. Amidst change there was constancy. Kitche Manitou heard songs, wailings and stories. He touched wind and rain. He felt love and hate, fear and courage, joy, sadness. Kitche Manitou meditated to understand his vision. In his vision, Kitche Manitou understood that his vision had to be fulfilled. Kitche Manitou was to bring into being and existence what he had seen, heard and felt. (1976, p. 12)

In Anishnaabe world view, then, Creation comes originally from *vision.* We, as people, come from the spirit world. In addition to generating living beings, the Creation process begins to lay out the key ideas and principles that constitute the foundation for our laws and codes of conduct: how we will relate to the Creator and all beings in Creation. These laws do not apply just to people but to all of Creation (sun, moon, stars, etc). The laws apply to Creation and we are simply part of it. The cycles described in this view of creation (e.g., birth, growth, and death) are all necessary for Creation to continue. In this world there are many *elements,* such as love, hate, fear, and courage. It is not a one-dimensional world. All things require acknowledgement, whether happy or sad, positive or negative. It is a world that strives for balance and harmony.

The laws that govern all beings of Creation and their relationships with each other, including people, come from the Creator. All beings in this version of Creation are to share its gifts. Each has equal entitlement. Kitche Manitou then goes about realizing his vision – the process of Creation: "Kitche Manitou then made the 'Great Laws of Nature' for the well being and harmony of all things and all creatures. The Great Laws governed the place and movement of the sun, moon, earth and stars; governed the power of the wind, water, fire, and rock; governed the rhythm and continuity of life, birth, growth and decay. All things lived and worked by these laws. Kitche Manitou had brought into existence his vision" (B. Johnston, 1976, p. 12).

Re-Creation

In the Re-Creation story, these teachings are reinforced. There has been a great flood and most of life on earth has perished, with the exception of birds and water creatures. Sky-woman survives and comes to rest on the back of a great turtle. She asks the water creatures to bring her soil from the bottom of the waters so that she may use it to make new land. The water animals (the beaver, the marten, the loon) all try to help her and fail. Finally, the muskrat volunteers, much to the scorn of the other water creatures. Although ridiculed, muskrat, the most humble of the water creatures, is determined to help. So he dives down, while the animals and sky-woman wait: "They waited for the muskrat to emerge as empty handed as they had done. Time passed. Smiles turned to worried frowns. The small hope that each had nurtured for the success of the muskrat turned into despair. When the waiting creatures had given up, the muskrat floated to the surface more dead than alive, but he clutched in his paws a small morsel of soil. Where the great had failed, the small succeeded" (B. Johnston, 1976, p. 12). There are many values and lessons to be learned from this story, but one of the most compelling is that all of Creation is important, all must be respected. If we lose or disrespect even the tiniest and seemingly most insignificant being, our own survival becomes threatened. In this story we learn of courage, sacrifice, and determination. We learn that all beings of Creation, including people, must work together to ensure the continuance of Creation. The Great Laws of Nature are reinforced. All beings of Creation are interdependent; we are related and we all have special gifts to contribute to the process of Creation. An important theme that emerges from this Re-Creation story and that is central to this chapter is the law that *people, too, must cooperate with all beings of Creation in order to survive.* Sustaining Creation requires cooperation and justice not among people only but among all beings of Creation.

In the Creation and Re-Creation stories, instructions are given by the Creator on how to relate appropriately with all beings of Creation. In the Anishnaabe world view, these instructions are often related to people in the form of stories, although there are other forms. Such instructions or knowledge is obtained by people from many sources, including Creation itself (B. Johnston, 2003; McGregor, 2004). Many stories and teachings are obtained from animals, plants, the moon, the stars, water, wind, and the spirit world. Knowledge is also gained through visions, ceremonies, prayers, intuitions, dreams, and personal experience. Basil Johnston (2003) writes: "Our ancestors learned what they knew directly from the plants, insects, birds and animals, the daily changes in the weather, the motion of the wind and the waters, and the complexion of the stars, the moon and the sun. They didn't write these down but kept them in their hearts" (p. vii). The stories tell the Anishnaabe that all our human and nonhuman relatives have roles and responsibilities that must be respected. Our stories also tell us that when balance and harmony are not respected among beings (including the spirit world), injustice will result and sustainability will be threatened. Teachings that emerge from Creation stories uphold ideas of holism and the importance of interrelationships among all elements of Creation. The earth is described as a living entity, bearing special responsibilities toward supporting the continuation of life.

Creation stories are fundamental to understanding the scope of environmental justice from an Anishnaabe point of view. We have to rethink what the terms *we* and *our* mean in this context. Environmental justice includes our relationships with each other, including all plants and animals, the sun, the moon, the stars, the Creator, and so on. It is necessary to move beyond the human-centred approach to one of understanding, accepting, enacting, respecting, and honouring relationships with all of Creation.

Indigenous Knowledge Framework

Traditionally, Anishnaabe people understood their relationship with Creation and assumed the responsibilities given to them by the Creator. The relationship with Creation and its beings was meant to be maintained and enhanced, and the knowledge that would ensure this was passed on for generations over thousands of years. The responsibilities assumed by individuals, communities, and nations as a result of having this knowledge ensured the continuation of Creation (what academics now refer to as "sustainability"). This knowledge I call Indigenous, or Aboriginal, knowledge. Indigenous knowledge can also be characterized as a *process*. As Anishnaabe storyteller Basil Johnston writes in his book *Honour Earth Mother* (2003), "Learning comes not only from books but from the earth and our surroundings as well . . . What our people know about life and living, good and evil, laws and purposes of insects, birds, animals and fish comes from the earth, the weather, the seasons and plants and other beings. The earth is our book . . . alive with events that occur over and over for our benefit. Mother Earth has formed our beliefs, attitudes, insights, outlooks, values and institutions" (p. v). LaDuke (1999) refers to the concept of "Minobimaatisiiwin," which means "the good life" and involves concepts of revival, rebirth, and renewal. Life is thus understood in terms of cycles and relationships within and among these cycles. A critical point in LaDuke's view is that, in order to understand Minobimaatisiiwin, and in order for the Aboriginal knowledge inherent in this way of life to have any real meaning, you must live it. In living such a life, people acted according to the tenets they perceived to be obtained from the Creator. Johnston (2003) continues: "What our ancestors learned of the land, the wind, the fire and the waters . . . is revelation, no less than is dream . . . Kitchi-Manitou shows us, speaks to us. Our ancestors watched and listened. The land was their book. The land has given us our understandings, beliefs, perceptions, laws, customs. It has bent and shaped our notions of human nature, conduct and the Great Laws. And our ancestors tried to abide by those laws" (p. xi). Aboriginal knowledge, then, comes from the Creator, from the earth, from Spirit, from our relationships with our ancestors. As Johnston (1976) puts it, these "truths must be lived out and become part of the being of a person" (p. 7). This is the "lifeway" that has sustained, and will continue to sustain, Anishnaabe nations (McGregor, 2004).

Environmental Justice and Water Quality in First Nations Communities

Just as the Creation and Re-Creation stories tell us, change is continually occurring; transformation and cycles are normal. Times are now very different from what they once were as we struggle to establish political relationships with the "newcomers" through seemingly endless processes of conflict and negotiation. Change, however, does not spell the end of Indigenous knowledge or of the requirement to respect traditional laws. What, then, can traditional Anishnaabe concepts bring to the current discourse on environmental justice?

An example currently facing Indigenous peoples all over the world relates to water. As such, Indigenous people have declared 22 March Indigenous World Water Day and have produced an Indigenous peoples' declaration on water (see www.indigenouswater.org).

In Ontario, numerous events over the past few years have generated much interest in protecting water quality. Of particular concern is the challenge of ensuring that First Nations receive the same measure of water quality protection as non-Native communities. However, non-Aboriginal communities have also begun to experience serious water quality issues. In 2000, the southern Ontario community of Walkerton saw seven people die from the contamination of the local drinking-water supply. This, of course, drew national attention to the seriousness of the problem of adequately protecting water. This same issue has been a serious environmental issue for First Nations since long before the events at Walkerton (Assembly of First Nations, 1993).

An inquiry was subsequently held into the Walkerton tragedy, with a mandate to address water quality concerns across the province. Following two years of study, including many community consultations and invited submissions, Commissioner Dennis O'Connor released a report containing 121 recommendations. One of the findings of this work was that First Nations do not enjoy the same level of protection as non-Aboriginal communities. The O'Connor report recognized that First Nations face serious problems in relation to water quality and that jurisdictional issues among federal, provincial, and First Nations governments present particular challenges in attempting to resolve such issues (O'Connor, 2002; Auditor

General, 2005). Additional issues raised in works by Weaver (1996), LaDuke (1999), and Gedicks (1994) around colonial history and ongoing institutionalized racism make resolving First Nations water quality concerns even more complex. The O'Connor report therefore called for cooperation among these governments in working out an approach to ensuring safe drinking water for all.

Walkerton served as a catalyst in mobilizing the Province of Ontario to initiate water protection legislation. In the fall of 2005, however, a drinking-water crisis in a remote First Nation community in northern Ontario drew further national attention. The First Nation community of Kashechewan had been under a boil water advisory for two years because of *E. coli* contamination of drinking water. In October of that year, the Ontario government declared a state of emergency at Kashechewan, and hundreds of people were evacuated from the community, with many others requiring medical attention. This event served as a further catalyst for bringing national attention to a long-standing problem in First Nations communities. The Government of Canada had previously established a First Nations Water Management Strategy (a five-year plan that began in 2003) aimed at improving the safety of water supplies in First Nations communities. Judging by the events at Kashechewan, however, this clearly was not enough. Within a few months, the minister for Indian and Northern Affairs Canada announced the Plan of Action for Drinking Water in First Nations Communities that included two noteworthy aspects: the Protocol for Safe Drinking Water in First Nations Communities and the commitment to appoint a panel of experts to advise on the appropriate regulatory framework required to ensure clean water in First Nations communities (Indian and Northern Affairs Canada, 2007).

The work of the experts panel is to date the only component of the action plan that has addressed the critical issue of Indigenous knowledge with respect to water quality. The panel, established in May 2006, was given a mandate to develop options for a regulatory framework for water quality in First Nations communities. The panel was charged with gathering information from a variety of sources, including public hearings and written submissions. Public hearings took place throughout the country in the summer of 2006 and in December of the same year, and the panel has since released its recommendations in a two-volume report. Although it remains to be seen what the final outcome will be in relation to how exactly water quality protection will occur in First Nations and their traditional territories, Indigenous knowledge does appear to be under consideration, albeit in a limited manner. For example, in the *Report of the Expert Panel on Safe Drinking Water for First Nations* (Swain, Louttit, and Hrudey, 2006), it is noted that "traditional attitudes toward water are holistic and spiritual" (p. 32). Although not explicit in recognizing Indigenous knowledge, the expert panel makes recommendations for the development of regulations that respect "customary law," recognized due to the "strong stewardship role for First Nations where water is concerned" (p. 57). Customary law is based on traditional world view, philosophy, principles, values, and knowledge.

This recognition of Indigenous knowledge remains limited, and there is as yet little if any guidance on how the consideration of Indigenous knowledge is actually to occur and who might be involved in such a process. It is increasingly obvious that input is required at the grassroots level from holders of Indigenous knowledge (Elders, traditional teachers, hunters, trappers, etc.) to answer the questions of the value and appropriate consideration of Indigenous knowledge in protecting water in these contexts. The following insights are gained from various initiatives (see Kamanga, Kahn, McGregor, Sherry, and Thornton, 2001; Lavalley, 2006; McGregor and Whitaker, 2001; Noojimawin Health Authority, 2006, for more detail) undertaken by the Chiefs of Ontario that offer a unique Indigenous knowledge perspective on appropriate relationships with water.

The Importance of Water

> Water is a sacred thing. This is reflected in many traditional beliefs, values and practices.
>
> – Elder Ann Wilson, quoted in McGregor and Whitaker, 2001, p. 17

The following pages express teachings and understandings shared by First Nations people who have participated in various water-related initiatives I have been involved with over the past eight years. My formal involvement in this topic began with work I undertook on traditional knowledge and water in preparation for the Chiefs of Ontario submission to the Walkerton Inquiry in 2000, along with workshops and presentations since, and a two-day Honouring Water Teachings workshop held in Garden River First Nation and which began on Indigenous World Water Day, 22 March 2007.

From a First Nations perspective, water quality is not just an environmental or ecological issue. One of the main features of Aboriginal knowledge, based on thousands of years of living sustainably with Creation, is its holism: the recognition that all aspects of Creation are interrelated. Thus, degradation of water quality directly impacts the people, with the effects permeating every aspect of their lives. It threatens their very survival. Aboriginal land-use activities and ways of life are still very much a part of First Nations peoples' lives today. Such ways of life and the values they support depend heavily upon healthy ecosystems, including clean water. It has been shown time and time again in the history of Aboriginal-non-Aboriginal relations in Canada that environmental destruction (of forests, lands, animals, waters) threatens the existence of First Nations peoples (Assembly of First Nations, 1993; Kassi, 1996; RCAP, 1996). Water quality, then, is not just an environmental concern; it is a matter of cultural survival (McGregor and Whitaker, 2001). Among Native peoples, water is recognized as the lifeblood of the earth (a living and conscious being). In turn, water is therefore the lifeblood of the people in numerous ways (physically, mentally/intellectually, spiritually, and emotionally). Water is integrally tied to the cultural survival of the people. First Nations activists who have formed alliances to advocate and "speak for the water" are at the same time resisting the genocide of their people (Fixico, 1998; Porter-Locklear, 2001; Ransom, 1999; Sam-Cromarty, 1996).

Water is, and always has been, viewed as precious by Indigenous people. Concern for water is not new in our communities. It hasn't just come about because of the pollution we face today. Water has always been, and continues to be, recognized as a fundamental life-giving force. Perhaps the most telling expression of this is the phrase "Water *is* life." Water is not just closely associated with life, or merely part of life, but rather water is life itself (McGregor and Whitaker, 2001). Benton-Banai (1988), states that "the Earth is said to be a woman. In this way it is understood that woman preceded man on the Earth. She is called Mother Earth because from her come all living things. Water is her life blood. It flows through her, nourishes her, and purifies her" (p. 2). Furthermore, water itself is understood to be alive. "Water is a relation," connecting us to all of Creation; "We humans co-exist with water, we have to care for the water in order for water to be clean" (Lavalley, 2006, p. 39). We need to respect and treat water as a relative, not a resource.

Water plays an important part in the spiritual life of the Anishnaabe people. It is an integral component of many ceremonies held to show respect for water and to assist with its life-giving force. These ceremonies are conducted as one way to maintain and remind the people of their intimate connection with water. For example, Akii Kwe, a group of Anishnaabe women from Bkejwanong Territory speaking for water, write in its position paper on water quality:

> We use the sacred water in our Purification Lodge, in ceremonies of healing, rites of passage, naming ceremonies and especially in women's ceremonies. At these times, the teachings are spoken to the water and then it is passed around from one to another in the circle to be shared . . . At the change of the seasons, a pilgrimage to the water is carried out in order to honour the Spirit of the Water. Our people have always understood that this sacred and powerful water gives life, and can take it away. (1998, p. 3)

To understand our relationship to water, we must look at the whole ecosystem. A holistic approach is required. We must look at the life that water supports (plants/medicines, animals, people, birds, etc.) and the life that supports water (e.g., the earth, the rain, the fish). Water has a role and a responsibility to fulfil, just as people do. We do not have the right to interfere with water's duties to the rest of Creation. Indigenous knowledge tells us that water is the blood of Mother Earth and that water itself is considered a living entity with just as much right to live as we have (McGregor and Whitaker, 2001).

Josephine Mandamin, Anishnaabe Elder and inspiration for the Mother Earth Water Walk, is currently leading a walk around the entire Great Lakes Basin to raise awareness of the importance of water and our responsibility for it. Between 2003 and 2008, Josephine walked around Lakes Superior, Huron, Michigan, Ontario, and Erie. Based on her knowledge and experiences, Josephine refers to water as having personality, stating that each body of water has its own (Mandamin, 2007). Edward Metatawabin, from the James Bay area in northeastern Ontario, echoed this same message at the Honouring Water Teachings workshop, presented by the Chiefs of Ontario, when he referred to water as a "little brother" who was present at the meeting convened with Elders to talk about water. Edward remarked that respecting water as we talk about it requires not referring to it in the third person. Water is here listening and we need to respect it (Metatawabin, 2007).

Water and Women

> Everyone has a responsibility to care for the water. Women, however, carry the responsibility to talk for the water.
>
> – Elder Ann Wilson, quoted in McGregor and Whitaker, 2001, p. 20

As water is a giver of life, women, also life givers, have a special relationship and responsibility to water (Lavalley, 2006; McGregor and Whitaker, 2001). The first environment for new life is within a woman's body, and water continues to be a crucial component throughout one's life. The recognition of women's role in creating life along with water means that women and water have a special bond. This bond is often expressed in ceremonies, where the role of Anishnaabe women is to speak for the water. As Akii Kwe explains, "In the water ceremony we make an offering to water, to acknowledge its life-giving forces and to pay respect. We have a responsibility to take care of the water and this ceremony reminds us to do it. Women bring forth life, the life of the people. Water brings forth life also, and we have a special role to play in this responsibility that we share with water" (in McGregor and Whitaker, 2001, p. 24). For many years, Indigenous women have noticed changes in water quality, particularly because they have a close and special relationship to the water. In the process of rediscovery, revitalization, and healing, the women of Bkejwanong Territory (Walpole Island, Ontario) have organized themselves to speak for the water. Akii

Kwe is an informally organized grassroots group of women speaking only from what they know. They began by protesting what was happening to the water, especially pollution contaminating the waters flowing around Bkejwanong Territory. The women decided to speak for the water and try to stop such actions. In a 1998 submission on water quality issues, Akii Kwe members stated that in Bkejwanong, nature provides the foundation of Anishnaabe culture and the ways in which the people conduct themselves (systems of governance). As part of this, the people have a responsibility to act on behalf of the water (Akii Kwe, 1998).

As introduced above, Josephine Mandamin is a Grandmother fulfilling a prophesy of a woman who will walk around the Great Lakes to remind people of their responsibility to water. Each walk begins in the spring with a water ceremony, feast, and celebration and is led by a Grandmother. Routinely covering distances of over a thousand kilometres, each walk has the goal of raising awareness about water and trying to change the perception of water from that of a resource to that of a sacred entity that must be treated as such. On these journeys the Grandmothers carry a vessel of water and an eagle staff, which they take turns carrying. The beauty of this effort is that it is led by women who are fulfilling their role and trying to engage as many people as they can in raising awareness of the spiritual and cultural significance of water. The walks have inspired Anishnaabe women in other communities to organize their own water walks.

Conclusion

As an Anishnaabe woman, I have a responsibility to speak for water and address equity issues in relation to water. Sharing my work over the years, advocating for the inclusion of the Anishnaabe perspective in various capacities, has been and continues to be part of my responsibility to all my relations and to Creation as a whole.

In this chapter, I hope I have offered insights into how concepts of environmental justice are understood from an Aboriginal perspective. I would never argue that environmental justice is *not* about injustice among peoples; however, it is certainly more than that. It is about injustices among all beings of Creation as well. It is our responsibility to ensure that our actions result in environmental justice and not the opposite.

It is not just people who have responsibilities, however. All beings have responsibilities to fulfil, and recognizing this contributes to a holistic understanding of justice. Our interference with other beings' ability to fulfil their responsibilities is an example of a great environmental injustice, an injustice to Creation. As an example, water is a living, spiritual being with its own responsibilities to fulfil. The sun and moon are also our relatives that in turn have their responsibilities. To restrict our discussion of environmental justice to relations among people results in a limited discourse.

Similarly, it can be argued that because of their intimate relationship with the land, any injustice to Aboriginal people is an environmental injustice to the extent that it impairs the ability of Aboriginal people to fulfil their responsibilities to Creation. Conversely, any injustice to the environment that impedes the ability of Creation to fulfil its duties to Aboriginal people is an injustice to Aboriginal people. Of course, this is true of all people: we cannot survive without an environment that fulfils our needs for survival. It is simply time for all peoples of the world to recognize this explicitly and to act accordingly.

Miigwetch
(Thank you)

Note

1 *Anishnaabe* is the noun used to denote a person from that culture, the culture itself, or an adjective used to describe things related to that culture (similarly, a "Canadian"; Canadian culture; from a Canadian perspective). "Anishnaabeg" and "Anishnaabek" are alternative spellings of the plural of *Anishnaabe*. *Anishnaabe* itself is also variously spelled "Anishnabe," "Anishinaabe," and even "Nishnaabe." There are reasons behind each of these spellings (for example, the Manitoulin dialect is undergoing the process of syncopation), and the debate on standardization is ongoing (Corbiere, 2007, 2008). For ease of reading, I use the spellings "Anishnaabe" and "Anishnaabeg."

References

Akii Kwe. 1998. *Minobimaatisiiwin: We Are to Care for Her.* Position paper from Women of Bkejwanong (Walpole Island) First Nation, ON.

Ashini, D. 1995. The Innu Struggle. In B. Hodgins and K. Cannon (Eds.), *On the Land: Confronting the Challenges to Aboriginal Self-Determination in Northern Quebec and Labrador* (pp. 29-42). Toronto: Betelgeuse Books.

Assembly of First Nations. 1993. Environment. In *Assembly of First Nations: Reclaiming Our Nationhood; Strengthening Our Heritage* (pp. 39-50). Ottawa: Assembly of First Nations.

Auditor General of Canada. 2005. Drinking Water in First Nations Communities. In *Report of the Commissioner of the Environment and Sustainable Development to the House of Commons* (Chapter 5). Ottawa: Office of the Auditor General of Canada.

Benton-Banai, E. 1988. *The Mishomis Book: The Voice of the Ojibway.* Hayward, WI: Indian Country Communications.

Borrows, J. 2002. *Recovering Canada: The Resurgence of Indigenous Law.* Toronto: University of Toronto Press.

Coon Come, M. 1995. Clearing the Smoke Screen. In B. Hodgins and K. Cannon (Eds.), *On the Land: Confronting the Challenges to Aboriginal Self-Determination in Northern Quebec and Labrador* (pp. 5-17). Toronto: Betelgeuse Books.

Corbiere, A. 2008. Personal correspondence (e-mail) from Alan Corbiere, operations manager and cultural/historical research, Ojibway Cultural Foundation, M'Chigeeng, ON. 2 December.

–. 2007. Personal correspondence (email) from Alan Corbiere, operations manager and cultural/historical researcher, Ojibway Cultural Foundation, M'Chigeeng, ON. 8 May.

Fixico, D. 1998. *The Invasion of Indian Country in the Twentieth Century: American Capitalism and Tribal Natural Resources.* Niwot, CO: University Press of Colorado.

Gedicks, A. 1994. *The New Resource Wars: Native and Environmental Struggles against Multinational Corporations.* Montreal: Black Rose Books.

Indian and Northern Affairs Canada. 2007. *Plan of Action for Drinking Water in First Nations Communities* (Progress report, 22 March 2007). Ottawa: Indian and Northern Affairs Canada.

Johnston, B. 2003. *Honour Earth Mother.* Wiarton, ON: Kegedonce Press.
–. 1976. *Ojibway Heritage.* Toronto: McClelland and Stewart.
Johnston, D. 2006. *Connecting People to Place: The Power and Relevance of Origin Stories.* (Unpublished lecture notes.) University of Toronto.
Kamanga, D., J. Kahn, D. McGregor, M. Sherry, and A. Thornton (Contributors). 2001. *Drinking Water in Ontario First Nation Communities: Present Challenges and Future Directions for On-Reserve Water Treatment in the Province of Ontario* (Submission to Part 2 of the Walkerton Inquiry Commission). Toronto: Chiefs of Ontario.
Kassi, N. 1996. A Legacy of Maldevelopment: Environmental Devastation in the Arctic. In J. Weaver (Ed.), *Defending Mother Earth: Native American Perspectives on Environmental Justice* (pp. 72-84). Maryknoll, NY: Orbis Books.
LaDuke, W. 2005. *Recovering the Sacred: The Power of Naming and Claiming.* Cambridge, MA: South End Press.
–. 1999. *All Our Relations: Native Struggles for Land and Life.* Cambridge, MA: South End Press.
–. 1994. Foreword: A Society Based on Conquest Cannot Be Sustained. In A. Gedicks (Ed.), *The New Resource Wars: Native and Environmental Struggles against Multinational Corporations* (pp. ix-xv). Montreal: Black Rose Books.
Lavalley, G. 2006. *Aboriginal Traditional Knowledge and Source Water Protection: First Nations' Views on Taking Care of Water.* Toronto: Chiefs of Ontario and Environment Canada.
Mandamin, J. 2007. *Women and Water.* Presentation at Honouring Water Teachings, workshop by Chiefs of Ontario, Toronto.
McGregor, D. 2004. Coming Full Circle: Indigenous Knowledge, Environment and Our Future. *American Indian Quarterly, 28*(3/4), 385-410.
McGregor, D., and S. Whitaker. 2001. *Water Quality in the Province of Ontario: An Aboriginal Knowledge Perspective.* Toronto: Chiefs of Ontario.
Means, R. 1996. Foreword. In J. Weaver (Ed.), *Defending Mother Earth: Native American Perspectives on Environmental Justice* (pp. xi-xiii). Maryknoll, NY: Orbis Books.
Metatawabin, E. 2007. Presentation at Honouring Water Teachings, workshop by Chiefs of Ontario, Toronto.
Noojimawin Health Authority. 2006. *Aboriginal Traditional Knowledge and Source Water Protection.* Paper presented at First Nations House workshop, University of Toronto, 25 March 2006.
O'Connor, D. 2002. *Part Two: Report of the Walkerton Inquiry; A Strategy for Safe Drinking Water.* Toronto: Ontario Ministry of the Attorney General.
Porter-Locklear, F. 2001. Water and Water Quality Issues in and for American Indian Communities. In K. James (Ed.), *Science and Native American Communities* (pp. 111-18). Lincoln: University of Nebraska Press.
Pushchak, R. 2002. Environmental Justice and EIS: Low-Level Military Flights in Canada. *International Journal of Public Administration, 5*(2), 169-91.
Ransom, J. 1999. The Waters. In *Haudenosaunee Environmental Task Force: Words That Come Before All Else: Environmental Philosophies of the Haudenosaunee* (pp. 25-43). Cornwall Island, ON: Native North American Traveling College.
RCAP (Royal Commission on Aboriginal Peoples). 1996. Lands and Resources. In *The Report of the Royal Commission on Aboriginal Peoples: Restructuring the Relationship* (Vol. 2) (pp. 421-685). Ottawa: Canada Communication Group.
Sam-Cromarty, M. 1996. Family Closeness: Will James Bay Be Only a Memory for My Grandchildren? In J. Weaver (Ed.), *Defending Mother Earth: Native American Perspectives on Environmental Justice* (pp. 99-106). Maryknoll, NY: Orbis Books.
Swain, H., S. Louttit, and S. Hrudey. 2006. *Report of the Expert Panel on Safe Drinking Water for First Nations.* (Vol. 1). Ottawa: Minister of Indian Affairs and Northern Development.
Washington Office of Environmental Justice. 1991. Principles of Environmental Justice. http://www.ejnet.org/ej/princples.html.
Weaver, J. (Ed.). 1996. *Defending Mother Earth: Native American Perspectives on Environmental Justice.* Maryknoll, NY: Orbis Books.

2
Reclaiming Ktaqamkuk: Land and Mi'kmaq Identity in Newfoundland

Bonita Lawrence

For Indigenous peoples, land is crucial to how identity is understood. For Indigenous peoples in Canada, being severed from their traditional land base by settler governments has frequently involved struggles against the government bodies created to "manage resources," and those agencies' representatives, such as provincial game wardens or fisheries control officers, who routinely prevent Aboriginal peoples from relying on the land for their sustenance. At the same time, Native peoples are prevented from protecting the land from those who turn it into "resources" to be consumed in mining, forestry, or hydroelectric development, devastating the land and making it almost unlivable. Meanwhile, the white-dominated environmental groups who seek to "protect" the land from development seldom manage to overcome the colonial logic that permeates their assumptions that this is *their* land to protect. And finally, this dynamic is sometimes brought to a close by policies that promote direct settlement on lands formerly thought of as "wild," so that Native people lose the last vestiges of control over their lands, which now become somebody's private property. These dynamics are, unfortunately, all-too-familiar in Native communities across Canada.

The situation of the Mi'kmaq of Newfoundland is, however, quite distinct. Most Newfoundland Mi'kmaq communities are not federally recognized under the Indian Act. Their way of life rooted in living off the land has been almost completely undermined, but because they are not federally recognized there have been no reserves to enable these communities to maintain a necessary distance from white settlers and to protect the language and maintain a more direct transmission of cultural knowledge within families. Without federal recognition and reserves, they also lose access to significant on-reserve funding that might provide them with alternatives to enable them to survive as cohesive communities. Many of the community leadership today were children when their parents, forced to engage in the wage economy, began to be silent about their identities and stopped speaking the language. Over the past thirty to forty years, as resource rape damaged the land and destroyed habitat for wildlife, and punitive hunting and fishing regulations were subsequently introduced, the generation who are now leaders was curtailed from engaging in any aspects of the way of life that had kept the communities together – the hunting, fishing, and gathering of berries and medicines that allowed for some degree of cultural continuity even if people had already lost the language and extensive cultural knowledge. Their children, the third generation shaped by intensive displacement from the land, are growing up in a society that ignores their presence, without knowing their language and without any connection to the practices of life that had sustained the culture on the land for so long.

The situation is rendered even more complex by the peculiarities of Newfoundland's history and by the widespread assertion, by Newfoundlanders and academic "experts" alike, that only the Beothuk, and not the Mi'kmaq, are Indigenous to Newfoundland. Meanwhile, the longer community members are forced to struggle with processes of assimilation and the legacy of cultural loss, the more difficult it becomes to acquire federal recognition as "Indians."

This chapter explores the relationships between Mi'kmaq people and the land, as expressed through interviews conducted in 2004-5 with the leadership and community members of a number of Mi'kmaq bands in southern, western, and central Newfoundland. It also explores the ways in which contemporary Mi'kmaq communities are engaged in cultural revitalization and asserting a sense of Mi'kmaq place in Newfoundland – a revitalization in which land rights and environmental protection are central. However, I will begin with some exploration of the background context in Newfoundland – the denial of federal recognition at the time of Confederation, and the historic denial, within Newfoundland, that Ktaqamkuk is part of the traditional Mi'kmaq homeland.[1]

Identity Legislation and the Erasure of a Newfoundland Mi'kmaq Presence

Since the mid-nineteenth century, Native communities, at the point at which they negotiate treaties or are in other ways severed from their land base, are brought under the Indian Act, a legal regime that not only controls every aspect of their lives but also that demands the reordering of their very identities – from being "the people," citizens of Indigenous nations, to being "Indians" under a colonial government. A crucial characteristic of legal regimes such as the Indian Act in Canada (or blood quantum/ federal recognition legislation in the United States) is the manner in which they create a grammar of racial difference, to divide Indigenous peoples through the creation of different categories of "Indianness" based on supposed degrees of "authenticity," whether it be of blood or of culture. A number of writers (Lawrence, 2004; Garroutte, 2003; Sturm, 2002; Monet and Skanu'u, 1992; Clifford 1988) have explored how such standards of authenticity are constantly being invoked in everyday understandings of indigeneity, to evaluate who should be considered to be Indian and who is not. This grammar of racial difference then shapes how "Indianness" is understood, not only among settler governments who deal with "Indians" and within the settler society in general but, even more crucially, among indigenous peoples themselves, fragmenting them and dividing their resistance to further colonial encroachment.

Paradoxically, in contemporary times, these legal regimes have become the means with which Indigenous peoples struggle with settler governments to redress land theft or treaty violations – and indeed, in many cases, to obtain the funding that ensures their very survival *as* communities. Furthermore, an unanticipated product of the legal and spatial segregation from white society that these legal regimes have created has been to provide some space, in reserve settings, for Indigenous linguistic and cultural survival (indeed, it was precisely to remove the protection that reserves have afforded Native cultural life that policies of genocide specifically focused on cultural destruction – residential schooling, forced adoption, and the outlawing of ceremonies – were implemented). But even with the outlawing of ceremonies and the advent of residential schooling, the isolated reserves set apart under the Indian Act as part of the reshaping of the identities of those thought to be "a dying race" have long been sites where Native people have maintained cultural and physiological distinctiveness from white people.

What happens, then, to Indigenous communities that are severed from their land and overwhelmed by settlers without federal recognition through legal regimes such as the Indian Act? How do such communities survive *as* Indigenous communities? What are the challenges they face? To ask this question is to consider the situation that Newfoundland Mi'kmaq people, who lived a relatively unrestricted life on the land until the early twentieth century, are faced with today, almost sixty years after Newfoundland joined Canada without bringing the Indigenous peoples of Newfoundland under the Indian Act.

It is important to recognize that the initial terms of union between Newfoundland and Canada, in 1947, involved Canada taking responsibility for Newfoundland Indians under the Indian Act. A year later, however, Canada denied that this term existed and stated that the Indian Act could not be implemented in Newfoundland because there was no term specifically addressing this (Hanrahan, 2003, p. 219). One possible reason for this is that Newfoundland joined Confederation at a time when Canada was already exploring possible ways of divesting of federal responsibility for

Indians. For example, between 1946 and 1949, a Special Committee of the Senate and the House of Commons on the Indian Act had recommended the assimilation of Canada's Indians and the removal of the status they had under the Indian Act (Hanrahan, 2003, p. 219). To deny Indian status to Newfoundland's Mi'kmaq, who at that point maintained relatively distinct societies but were already facing growing settler encroachment and had no reserves to retreat to, would be the perfect opportunity for the federal government to observe how long Native people could hang on to their identities without either land or legal status as Indians. Indeed, a report by Canadian human rights commissioner Noel Lyon speaks to the success of this experiment when he notes that throughout the twentieth century, hundreds, if not thousands, of persons of Mi'kmaq ancestry have been assimilated into the larger society surrounding them or have left Newfoundland (Lyon, 1993, p. 9).

The profound injustice of this can perhaps be most clearly articulated by examining the circumstances of Mi'kmaq people in different parts of the Atlantic provinces. The Mi'kmaq of New Brunswick, Nova Scotia, and Prince Edward Island, as well as parts of Quebec, came under federal jurisdiction in 1867. Curtailed from an active life on the land for far longer than Newfoundland Mi'kmaq, they nevertheless, in reserve communities, have been able to maintain the Mi'kmaq language and a distinct identity as Mi'kmaq, without ever having to "prove" their Indianness. Their relatives in Newfoundland, however, having been denied federal recognition in 1949 while they were still land-based, and being subsequently forced off the land in conditions that enforced a silence about identity and culture, now have to "prove" their Indianness to be recognized as Indians, which generally requires that there be clear and unequivocal distinctions between Native people and non-Natives (Lawrence, 2004, p. 4).

By the 1970s, Newfoundland Mi'kmaq began increasingly to organize to address their claims for recognition. By 1981, the Ktaqamkuk Ilnui Saqimawoutie (KIS) and the Conne River Indian Band Council, together representing approximately eighteen hundred Newfoundland Mi'kmaq, submitted a comprehensive land claim to about the southern third of the island of Newfoundland, which encompassed the traditional territory that Mi'kmaq people had always used (Bartels, 1987, p. 32). The Canadian government has, numerous times, alternately entered into negotiations and forced the matter to go through the courts. The Newfoundland government, meanwhile, has consistently refused to cooperate. In 1985, when genealogical research established that the ten bands then involved with the Federation of Newfoundland Indians were qualified to begin the registration process under the Indian Act, the Province declined to enter into the negotiations (Lyon, 1993, p. 6). Furthermore, in 1995, when Canada was funding the Conne River band in its research to establish a land claim,

Newfoundland because the first province in Canada to unequivocally reject a land claim. Then-premier Brian Peckford stated that Newfoundland would not negotiate a land claim with the Mi'kmaq because their claim to Aboriginal land use and occupancy of the south coast of Newfoundland had not been substantiated (Hanrahan, 2003, p. 228). In this context, it is worth noting that for over one hundred years, the category of "Indians" or "Micmac Indians" appeared in the Newfoundland census, suggesting that the government of Newfoundland has long been aware of a Mi'kmaq presence throughout the island. Indeed, Newfoundland had a law making it illegal to sell liquor to Indians, and in 1945, a census of Newfoundland Native people indicated 527 "Halfbreeds" and 431 Indians (Hanrahan, 2003, p. 229).

The registration of the people of Conne River, in Bay d'Espoir, under the Indian Act took place in 1984, after an extensive struggle of Mi'kmaq people across the province, including occupations of government offices and a hunger strike (Hanrahan, 2003, p. 221). It was clear that registration for Conne River was considered a possibility because it was an isolated community without significant European presence, and where a reserve had been surveyed in the late nineteenth century (Jackson, 1993, p. 116). But undoubtedly, the dynamics of allowing *one* reserve to be created (to become, almost by definition, the "real" Mi'kmaq of Newfoundland) is a crucial means through which colonial governments can deny the authenticity of all other Newfoundland Mi'kmaq communities. To successfully maintain a system of identity legislation, with its grammar of racial difference, the presence of *some* "authentic Indians" is the tool with which other Native people can be dismissed as inauthentic.

The leadership of the still-unrecognized bands affiliated with the Federation of Newfoundland Indians has continued to struggle, for over thirty years now, to be recognized as Indians. To keep a system of racial division intact, the condition imposed most recently by the federal government has been that the estimated 4,500 Mi'kmaq currently living in Newfoundland, those who are deemed to be eligible as "founding members," would be registered as members of a single landless band under the Indian Act, with no reserve base but with entitlement to certain federal programs (Sheppard, 2006). The other "choice," suggested by the Canadian Human Rights Commission, is that registration under the Indian Act is unviable for communities that share territory with non-Natives, and that therefore self-government agreements should be negotiated with the various communities of the Federation of Newfoundland Indians, involving options to preserve culture (Lyon, 1993, pp. 21-24). Nowhere is the question of land being addressed, even though most of the Mi'kmaq communities depended entirely on the land for their physical survival as recently as the 1960s, and their cultural survival still depends on it.

Across Newfoundland, it is precisely the relationship between Mi'kmaq peoples and the land that has been denied. A primary means of this has been through what is known as the Mi'kmaq mercenary myth – the notion that only the Beothuk, now extinct as a people, were truly Indigenous to Newfoundland and that the Mi'kmaq had been imported from the Maritimes as mercenaries, by the French, to destroy them. Newfoundland Mi'kmaq, then, have been presented to the public as just another settler group – somewhat like the Irish, who are not Indigenous to Newfoundland – and one, moreover, that is historically tainted by having murdered the Beothuk. In this version of history, the Mi'kmaq, as relative newcomers to Newfoundland, historically tainted with murder, are portrayed as "savage and uncivilized," distinct from the "civilized" Newfoundlanders of English or Irish descent and from the extinct but noble Beothuk (Bartels and Bartels, 2005, p. 10).

Increasingly, however, as succeeding generations of Mi'kmaq people are denied access to the land that was their primary cultural reference, white Newfoundlanders have another "basis" for which to deny the land rights of the Mi'kmaq people in their midst – simply the denial of their Indianness altogether. The grammar of racial difference, which separates "real" Indians from those Native people considered racially or culturally "inauthentic," is used in everyday ways to deny the validity of Mi'kmaq claims to be Indigenous people at all. Meanwhile, at all levels of Newfoundland society, there seems to be a habitual silence about Mi'kmaq presence. Most contemporary writers appear to have rejected the Mi'kmaq mercenary myth; however, they appear to have gone no further in accepting Mi'kmaq presence as a valid Indigenous presence.

It may well be that for Newfoundlanders, in some important way, national identity depends on the notion that the Beothuk are the only Aboriginal inhabitants of Newfoundland, precisely *because* they are extinct as a distinct people. To claim that the only Indigenous people on the island are extinct is to make contemporary Newfoundlanders unequivocally heirs to the land. And this appears to be important for several reasons. In every index – economic, social, political – since its earliest days, Newfoundland has remained suspended somewhere between Third World colonial conditions and the more privileged status of the white settler states. From the seventieth century, when a local colony finally began to take root in Newfoundland, most efforts to diversify its economy from the fishery have met with disaster. From the eighteenth-century plundering of Newfoundland's rivers, furs, seals, and birds, which devastated the Beothuk's food supplies and made life unsustainable for them, to nineteenth-century failed attempts at establishing farm colonies and building railroads to twentieth-century development of pulp mills and mining through heavy foreign investment (Alexander, 1980, pp. 26, 29), the colony remained resolutely shaped by the fishery and the poverty of most of its fishermen. When 47 percent of Newfoundland volunteers for the First World War and 57 percent of conscripts were rejected as medically unfit (Overton, 1980, pp. 81-82), the extent of poverty in Newfoundland became obvious. By 1932, when 25 percent of the population was on poor relief, and 60 percent of Newfoundland's income was consumed in servicing the interest on its national debt (Long, 1999, p. 117), Newfoundland surrendered its independence back to Britain and was ruled by a governor and six commissioners, from 1933 until 1949.

Confederation may have provided a slightly better standard of living for most Newfoundlanders and enabled a basic infrastructure of roads, hospitals, and schools to be built, but this has come at a terrible cost. The richest fishing area in the world, which sustained a global fishery for five centuries, was plundered and ruined in less than fifty years. Oil development revenues are currently being bled out of the province, and a sustained lack of apparent interest on the part of Canada in addressing any other form of economic development has meant that there has been a 12 percent (70,000 people) decline in Newfoundland's population between 1991 and 2001 because of out-migration (Royal Commission, p. 35). In this context, where Newfoundland remains the poorest and most isolated region of Canada, contemporary Newfoundland identity seems to rest on the notion that Newfoundland is Newfoundlanders' land. To admit the Mi'kmaq as being Indigenous to that land would be to risk decentring the strong nationalist claims of the non-Native people of Newfoundland. Many Newfoundland Mi'kmaq have been reluctant to drive a wedge between themselves and their non-Mi'kmaq neighbours by speaking about Mi'kmaq land rights. But the resulting silence has made it easier for notions such as "landless band" status to seem, at least on the surface, somewhat acceptable.

In 2004, interviews were conducted with the leadership of nine of the bands that are members of the Federation of Newfoundland Indians (FNI), as well as members of women's groups and the youth coordinator for the FNI. In 2005, further interviews were conducted with bands outside the FNI, such as the Ktaqamkuk Mi'kmaq Alliance and the Kitpu First Nation, as well as one person from Miawpukek (Conne River First Nation). Due to the distances involved and the resulting delays and difficulties in establishing connections with the leadership of the St. Albans band, on the northern peninsula, the St. Albans band was the only community in which the leadership was not interviewed. Amazingly, in view of the utter marginalization of Newfoundland Mi'kmaq and the range of ideological attacks that have been aimed at their very existence, all of the leaders that were interviewed have shown tremendous clarity about refusing the logic of racial divisions that Canada has imposed in recognizing one band but not all the others. This is extremely important. Across North America, wherever the grammar of racial division is been accepted, it has proved to be absolutely fatal for Indigenous empowerment. The recognition of Conne River, and not the other bands on the island, could have potentially been disastrous for Mi'kmaq empowerment generally. However, for the most part, the leadership continues to insist that, despite federal government divisions, all Mi'kmaq people on the islands – status and nonstatus – are one people. Therein lies the hope for Mi'kmaq people in Newfoundland – that as Mi'kmaq people they have another vision of what being Mi'kmaq is, one that does not rely on having, or lacking, Indian status; a vision that is traditional, relating to their having stayed on the land much longer than any other Native people in eastern Canada.

Land, Language, and Mi'kmaq Identity

> I have heard elders explain that the language changed as we moved and spread over the land through time. My own father told me that it was the land that changed the language because there is a special knowledge in each different place. All my elders say that it is land that holds all knowledge of life and death and is a constant teacher. It is said in Okanagan that the land constantly speaks. It is constantly communicating. Not to learn its language is to die. We survived and thrived by listening intently to its teachings – to its language – and then inventing human words to retell its stories to our succeeding generations . . . In this sense, all Indigenous peoples' languages are generated by a precise geography and arise from it.
>
> – Jeanette Armstrong,
> *Land Speaking*, p. 178

As Okanagan writer Jeannette Armstrong eloquently expresses, Indigenous peoples' cultural survival is deeply embedded in their relationship to land. For every Indigenous society, land is imbued with life, and although every part of the land (and the web of life that is anchored to it) is sacred, some parts of the land are recognized as sites of great power, with which spiritual leaders learn to mediate through ceremony. Many Elders have said that it is our connections to land, and the life that lives on it, which make us human – to lose those connections is to become dehumanized and lost, while to maintain connections to land is to remain whole, not only as individuals but as peoples (see Anderson, 2000; Silko, 1996; Walters, 1992). Indeed, for Indigenous communities in particular, the health of the individual is firmly tied to the health of the community as a whole.

For Ktaqamkukewaq Mi'kmaq (the Mi'kmaq people of Newfoundland), community bonds have been weakened through loss of access to the land. As a result, it is often necessary to consider in more formal ways what was once taken for granted – that they have been bound to the places of their ancestors since time immemorial. In a sense, it is important to understand the longevity of ties that Mi'kmaq people have to the land in order to appreciate what happens to cultural wholeness when people are removed. It also tells a different story of relations among the Indigenous peoples of the territory now known as Newfoundland.

Sakej Henderson, in consultation with Mi'kmaq *putus* (storykeepers), has conceptualized the relationship between Mi'kmaq peoples and Mi'kmaki, the "land of friendship," as a sacred order, where a shared, coherent, and interrelated world view existed, where the life forces of stones, trees, rivers, coasts, oceans, animals, and people resided in harmony (Henderson, 1997, p. 32). The world view of the Mi'kmaq flows from a Creation story, which not only moves seamlessly from mythical time into historical time around the end of the last ice age but also carries us to the sacred practices and relationships of the present (Battiste, in Henderson, 1997, pp. 13-19). Henderson has written how the Ancient ones followed the caribou back to Mi'kmaki as the glaciers retreated, about eleven thousand years ago. Indeed, Mi'kmaq oral tradition in Newfoundland recorded by Frank Speck (1982, p. 26) demonstrates that in the 1920s, Mi'kmaq people still spoke of the Sa'yewedjkik, the "ancient ones" who occupied Newfoundland and other Mi'kmaq territories long ago. These "ancient ones" were possibly the ancestors of both the Mi'kmaq and Beothuk. DNA examination of Beothuk remains is being examined to determine whether, among other things, the Mi'kmaq and Beothuk have common ancestry (Calamai, 2005, p. 1). By about 5000 B.C., a network of Ni'kmaq (kin friends) had formed. They travelled extensively in their ocean-going canoes, engaging in a trade network that ranged from northern Newfoundland down as far as Central America. By the tenth century A.D., large numbers of Ni'kmaq had formed into a confederacy of allied people, known among themselves as "ulnoo" (the people) but confederated as the Mi'kmaq nation, with territories encompassing six regions of what is now the Atlantic provinces. Ktaqamkuk, the "land across the water" from Cape Breton, was part of Unamakik (the foggy land), the first territory where the grand chief lived.[2] Although the distances between these parts of the territory were extreme, Speck was told how, even in the early twentieth century, Mi'kmaq in their ocean-going canoes still travelled the ninety-three miles from Cape North (in Cape Breton) to Cape Ray, Newfoundland, in two stages, with a stop at St. Paul's Island. As his informants indicated, when a party wished to cross, the most experienced men travelled over first, as quickly as possible. Once they had landed, the greater number of the travelling party followed at night, when the waves were calmer, guided by a beacon on the high barrens of Cape Ray kindled by the first group (Speck, 1982, p. 26).

Broader geopolitical allegiances were developed at the time of the formation of the Mi'kmaq nation. First came the alliances of allies from the other Atlantic nations as the Wabanaki Confederacy, the people of the Dawn. These included the Beothuk (upriver people) in Newfoundland, the Wulustukw keuwiuk (beautiful river people), the Maliceet-Passamaquoddy of southeastern New Brunswick and northeastern Maine, and the Eastern Abenaki of Maine (Henderson, 1997, p. 32). The Wabanaki Confederacy also formalized relationships with other peoples further afield, such as the Iroquois Confederacy and the Anishnaabe Confederacy of the Great Lakes (Battiste, in Henderson, 1997, p. 17), as well as a number of Innu groups on the north shore of the St. Lawrence and in Labrador, Inuit in the Strait of Belle Isle, and, in the 1500s, the St. Lawrence Haudenosaunee. When treaties were negotiated with the Haudenosaunee, a seventh region, the Gaspé became part of the Mi'kmaq nation (Henderson, 1997, pp. 16-17). With the great changes that colonization brought to the land, Ktaqamkuk became a separate region from Unamakik, forming the eighth district of Mi'kmaki in 1863 (Benwah, 2005b).

According to the Elders, Mi'kmaki is "owned" in a formal sense only by unborn children in the invisible sacred realm (Henderson, 1997, p. 32); however, its regions are also traditionally governed by a grand council, or Mawiomi. At still another level, in an effort to resist invasion, in 1610 the Mawiomi negotiated a concordat that consolidated Mi'kmaki formally as a Catholic republic under Rome (p. 87). Whatever the conditions of contemporary Mi'kmaq communities, the reality remains that all of these spiritual and geopolitical relations, past and present, connect Mi'kmaq people, including Ktaqamkukewaq Mi'kmaq, with Mi'kmaki; these stories are the histories that sustain Mi'kmaq presence, and that contemporary "experts" must routinely ignore in order to continue to deny Mi'kmaq historic presence in Newfoundland.

The Colonial Order in Mi'kmaki

The advent of colonization brought different kinds of colonial policies and processes to different regions of Mi'kmaki. In what is now the Maritime provinces, extreme colonial repression under the British was followed by extensive settlement and Canada's imposition of the Indian Act and the reserve system. By comparison, Newfoundland, a fishing colony of Britain since the sixteenth century, experienced only sporadic attempts at settlement throughout the seventeenth and eighteenth centuries, primarily along its coastlines on the eastern and northern side. Newfoundland's nineteenth-century attempts at independent nation building were largely unsuccessful; consequently, it surrendered its independence back to Britain until 1949. Because of these different policies and processes of colonization, no formal legal regime regulating Native identity, such as the Indian Act, was imposed in Newfoundland; moreover, until well into the nineteenth century, the greater part of its white population remained established primarily along the shorelines, with areas of resource development maintained in specific locations within the interior. Because of this, Mi'kmaq people in Newfoundland enjoyed a relative freedom and access to land within the interior and along the west coast long after Maritime Mi'kmaq had been subjugated and reduced to tiny reserves, the bulk of their land and livelihood taken. For example, in 1822, at a time when the cumulative effect of bounty hunting for Mi'kmaq scalps, "scorched earth" policies, starvation, and the resulting epidemics meant that Maritime Mi'kmaq were in danger of becoming extinct, explorers in the interior of Newfoundland spoke with Mi'kmaq who had never seen a white man before. Until the late 1800s, Newfoundland Mi'kmaq travelled all over the land, living and utilizing the resources, from St. Mary's and Placentia Bay in the east to Notre Dame and Bonavista Bays in the north and along the entire west coast, using extensive overland routes throughout the interior (Hanrahan, 2003, p. 207).

This is not to suggest that the interior of Newfoundland was free of white presence. But for many years, the Beothuk bore the brunt of white encroachment, including the plunder of their food supplies, simply because the Europeans were invading their territories and not those of the Mi'kmaq. Mi'kmaq territory has always been the south and west coasts, and the interior south of Red Indian Lake. The Beothuk appear to have relied extensively on the Trinity Bay area of the Avalon Peninsula for sealing and gathering birds' eggs, as well as on Funk Island and the shores of the Bonavista Bay, particularly as European presence on the Avalon Peninsula drove them further inland and north. They also relied extensively on the rich salmon rivers of the northern interior, as well as the caribou migrating down from the northern peninsula at the Exploits River and Red Indian Lake (Marshall, 1998, pp. 63-78). Until the Europeans increasingly invaded Beothuk territory, this division of territory appears to have worked well for both groups. Indeed, the oral tradition of the Mi'kmaq suggests that the Beothuk and the Mi'kmaq shared a village in St. George's Bay and that there was a good deal of intermarriage between the two nations (Hanrahan, 2003, p. 207).

By comparison with the Beothuk, the Mi'kmaq faced much less pressure from white encroachment during the eighteenth century. In general, the British reluctance to fish along the south coast west of Burin Peninsula was because of Mi'kmaq presence; this was the last area of Newfoundland to be settled by Europeans (Hanrahan, 2003, p. 206). On the west coast, where the British had permitted France to fish along what became known as "the

French shore," a number of Mi'kmaq had absorbed French fishermen into their families; as a result, on parts of the west coast, Mi'kmaq language became interspersed with French. However, offsetting the French influence was a steady migration of Maritime Mi'kmaq to Newfoundland throughout the eighteenth and nineteenth centuries. It is perhaps not surprising that in the face of scalping proclamations and policies of deliberate starvation, and the withholding of land for reserves for almost a century after 1751, so many Maritime Mi'kmaq, particularly from Eskasoni on Cape Breton but from other communities as well, would wish to join their relations in Newfoundland.

By the mid-nineteenth century, when Newfoundland had gained some level of local self-government, Mi'kmaq people began to feel the colonial presence more intrusively, as policies to reduce dependence on the fishery and develop the interior gradually began to increase the spread of settlers into Mi'kmaq territories. Through interviews with the leadership of a number of Newfoundland bands, a picture of Mi'kmaq settlements in the mid-nineteenth century, the time when white people were just beginning to encroach, begins to emerge.

Calvin White, of Flat Bay, recalls how the people of his community, formerly clustered around St. George's Bay, the traditional heartland of Mi'kmaq presence in Newfoundland, retreated to communities such as Flat Bay and Conne River in the face of white encroachment:

> The history of our community that I can recollect from the teachings goes back quite a number of years, probably back to some time in the 1600s. The community of Flat Bay is not a particular land-base community, Flat Bay is the entire harbour inside the point of St George's Sandy Point – that is the bay. Actually, when people talk about Flat Bay dating back in the 1700s and 1800s they are talking about the geographic area of the harbour itself that is Flat Bay and the surrounding communities. The Indian people surrounding these communities are part of that era, but the community came into being as the European population started to infiltrate the community, that was then occupied by Mi'kmaq people. Some of the people moved to Conne River . . . and some of the people moved into Flat Bay, and other families even remained in the main part of the St. George area. There are still families that remain there today but because of the pressures by the European communities most of the Indian people moved to these locations, like Conne River and Flat Bay.

Along the south coast, in Burgeo, Bernard Benoit, of the Caribou band describes how his ancestors lived in the mid-eighteenth century, just before white people began to settle in the area:

> The people in Burgeo primarily come from two brothers, two Benoit brothers, who lived in White Bear Bay, in around the 1860s. They were born, I believe, around White Bear Bay in 1848. There was a Lawrence Benoit, and a Peter Benoit, two brothers. They came up the coast to a little place called Bay Lou and they married a couple of the white women that were there, and that is where the Indian population of Burgeo come from now. It can go back directly to those two brothers from White Bear Bay, Lawrence Benoit and Peter Benoit. Of course, there were their brothers, Frank Benoit and George Benoit, and they were Conne River People . . . The Benoits, the Joes, and the Stephens intermarried and that is where my people come from. Of course we extend pretty much all the way down to Conne River. So, down on Conne River you got two brothers, and two of the brothers were up here . . . They had tilts. You leave Burgeo here, and you go in ten miles and you have a little trapper's tilt, so every ten miles was a tilt. My grandfather, when he would leave here in his trapping days, he would leave here with 120 pounds on his body. He would have a 100-pound sack of flour, plus he would have a few things, and his buddy would have about 100 pounds of tar felt, and traps. They would go ten miles the first day to the first tilt. They would drop off a little bit of felt, do the repairs, and leave so much flour. The first trip would be to stock all of the places. They would go right up to Red Indian Lake.

Changes to the Mi'kmaq way of life began in the late 1800s as industrial development on their land, particularly hydroelectric development and large-scale logging, made life dependent entirely on the land much more precarious (Hanrahan, 2003, p. 209). Policies of settlement in the interior were also implemented – from 1873 through 1884, legislation was passed to offer free land to settlers willing to clear it and to promote agriculture as well as mining, forestry, and homesteading. Newfoundland was also surveyed into townships, sections, and quarter sections around this time (Alexander, 1980, p. 26). While the impetus to develop Newfoundland economically was supposed to be about self-sufficiency and economic development, in fact, foreign investment built the railroad, dry docks, mining, paper mills, coal mines, telecommunications networks, oilfields, and all significant manufacturing. Each investment agreement provided concessions in the form of Crown land grants, drawbacks on duties, tax holidays, and protective tariffs whenever possible, with no provisions for Newfoundland equity participation (pp. 29-33).

With these developments, it was Mi'kmaq people who paid the price in loss of land and livelihood, as white settlers began moving further into areas that had traditionally been occupied primarily by Mi'kmaq. Below, Bernard Benoit describes how this was experienced in the Burgeo area:

> From 1850s to the 1950s, if you take that period of time, then the Native people that lived around Burgeo were living out of Grandy's Brook, they were based there. My great-grandfather Francis Benoit had a farm there, he used to farm and raise cattle, pigs, chickens, and things like that. He would do a little bit of farming, and bring it down and sell it in Burgeo to the stores – that would have been back in the late 1800s. Basically, Burgeo got started in 1795 by the first white settlers. By the middle of the 1800s, of course, the population was expanding, growing, and they were moving into Grandy's Brook. It is my understanding, although I am trying to find more about the history of this place, that most of those Indian peoples moved back into the country and eventually into the Conne River area because their traplines would run from Grandy's, to Conne River and those places. With the coming of the white people, the Indians sort of moved back. I think all of the families moved back, with the exception of the Benoit family because my grandfather had a farm there and was sort of tied to the land . . . I think that most of the Native people, that is where they were but, of course, I guess the story from the 1900s to the 1930s is the Indian influence here died out and disappeared. I guess this got to be a place where if you wanted to make a living, if you wanted to live in this society, then there was really only room for English people, there was no room for Indians.

Another development of the time was a successive series of game laws, passed between 1845 and 1896, to protect wildfowl and wild game. The earlier laws acknowledged that "poor settlers" needed to be able to hunt for food; however, by the early 1900s, all reference to hunting or fishing for food had vanished, and the gaming laws defined wildlife resources as recreational or sporting resources (McGrath, 1994, p. 211). The Department of Marine and Fisheries was created in 1898, and a Game and Inland Fisheries Act followed in 1910, creating a Game Board responsible for game protection and propagation until 1934, when a commissioner for natural resources was given the responsibility for making regulations, and the Newfoundland Ranger force was created, largely to enforce game and forestry laws (pp. 211-12). By this time, sportsmen's organizations were active in Newfoundland. The promotion of wildlife resources for tourism was part of the reason for building the railroad. Indeed, the Reid Newfoundland Company, which operated both the railroad and the steamships to the island, brought many hunters to the interior. They also operated hotels, to bring fishermen to the rich salmon rivers of the west coast (pp. 212-14). Although some Native men were hired as guides, the reality is that Mi'kmaq communities got squeezed out by the process.

With the wildlife protection regulations, Mi'kmaq began to face the same pressures that the Beothuk had experienced earlier as the rivers where they fished traditionally were appropriated by settlers, and the banning of the use of spears and weirs made it more difficult for Mi'kmaq to fish (Jackson, 1993, p. 161). These regulations also curtailed access to caribou; however, at least in the interior, most hunters were well beyond the reach of game wardens. Because of this, trapping for furs, when prices were still high, and sporadic logging sustained life on the land well into the beginning of the twentieth century.

From the 1920s to the 1940s, however, the fur markets declined steadily, until many Mi'kmaq were forced to turn to logging on a more sustained basis (Jackson, 1993, p. 165). The 1920s and 1930s saw another displacement of west coast Mi'kmaq communities as French fishermen, after the French shore was abandoned in 1904, were relocated to Port au Port Peninsula. Munegooskek (Red Island), off Port au Port, had been a French fishing station since the 1700s, and indeed, it was individuals from this group who had married into Mi'kmaq families from St. George's much earlier. A handful of other French from Red Island had settled in the Mainland area, Port au Port, but the resettlement of over a hundred French fishermen onto Mi'kmaq land in the area displaced a number of Mi'kmaq families (Benwah, 2005a).

With the bankrupting of the Newfoundland economy in 1933, Newfoundland was forced to surrender its dominion status and accept the control of six commissioners and a governor, four of whom were from Britain. The commission government remained in place until Confederation, in 1949. At that point, about one-third of the Newfoundland population was dependent on the fishery. Only a sparse network of local roads existed outside the Avalon Peninsula (Parker, 1950, pp. 82-83). Only about 7 percent of Newfoundland children attended elementary school; a large percentage was illiterate (pp. 120-22). In 1940, Newfoundland had the highest infant mortality rates (one in seven) in the "developed" world. There were no hospitals outside St. John's and negligible health care elsewhere (pp. 132-33).

During this interval, the circumstances facing Newfoundland Mi'kmaq are entirely undocumented. It is as if they did not exist. Indeed, in a turn that boosted the economy and raised the standards of living for many white settlers on the west coast, an American military base was built at Stephenville in 1941, even though this involved destroying a Mi'kmaq community. Linda Wells describes the effect on the community of Indian Head:

> In the area we live in here, because what happened to the traditional lands that were taken, I do not think there is any way that we can ever go back to the way we were. You know, like, you cannot put Indian Head

back together that they have blasted out. The things that were taken, are taken. They are gone. There are some things that we have to learn to live with, but I think some accommodation should be made, should the Aboriginal people in Bay St. George decide that they wanted powwow grounds or they wanted a designated area that would be an Aboriginal area. I feel that should have been part of what was negotiated because in our area here . . . the land has been so commercialized that there is only small pockets of land left, really, where we could actually say, "Designate this land for us . . ."

I can only speak for our area here, and really the only area that I can speak to is Indian Head. Traditionally, the village that existed here . . . and in the Main Gut area too . . . the village that existed at Indian Head when the Americans decided they were gonna have a base in Stephenville and they wanted a port, you know – they put the people off of their land and that should never have been done. They destroyed the mountain with the blasts. I know they looked at it as progress, and that it is the two-edged sword, because it helped a lot of people. But all that area was for traditional Mi'kmaq people. You know the port is here, where the ships go in, well over that section, the white that you see in that area was blasted and trenched. That was all Mi'kmaq land . . .

We need things to help us survive here. I hear people say on the radio, "Why do they need somebody to tell them they are Indian?" We do not need somebody to tell us that we are Indian – we know we are Indian. We want what was taken from us, what is ours by right. They came in here and they took, and took, and took, and they are still taking.

The more that Mi'kmaq people were forced out of a subsistence lifestyle and forced to rely on employment for survival, the greater was their vulnerability to racism. It was at this point that many Mi'kmaq began concerted efforts to suppress their cultural identity. Bernard Benoit succinctly explains what ensued: "There were no written laws, but I think that it was the practice of the day. If you wanted work, you had to give up in a way and you become English, and that was basically their way . . . So if you wanted to survive . . . you shut up about it. That was the case until the late 1980s." Although the more geographically isolated communities found it easier to maintain cultural cohesion through their distance from white people, other communities located close to areas of industrial development faced an increasing presence of white settlers, and pressures, for the first time, to reduce signs of difference between them and the surroundings whites. Suppression of language, and eventually a tremendous silence even about openly identifying as Mi'kmaq, began to be a factor in some communities, the degree depending on the extent to which white people dominated their environment. Ignatius Paul (known as "Nish"), from Badger, an Elder of the Exploits band who is now deceased, described his experiences as a child as his parents stop speaking the language in an attempt to protect the family from racism – which, he believed, was in vain:

> Well, you know, I remember my father and them, when I was a kid, you know, they used to sing the songs, right? Mi'kmaq songs, right? But you know, we never learned it – they stopped it, they stopped speaking their language when they started working for these companies, and stuff like that, right – as a matter of fact, my grandfather, in those days, they were even hiding at that time . . . Well, I guess everybody feels pretty much the same, you know what I mean? You know, their identity was gone, but it's coming back again, right? You know, when I was growing up in Badger, as a boy, there was people's houses that I wasn't allowed to go into, right? I wasn't allowed to go into their houses – I could go there and split their wood and stuff like that. But I wasn't allowed to go in through their doors. They'd bring me out a glass of syrup and a piece of cake, and I'd have to sit on the chopping block and eat it there. I wasn't allowed inside their doors . . . I lived through that, right? You know, with a broken heart that nobody knew anything about, you know – I mean, even if I started school, you know, I went out on parties and stuff like that and I just sat in the corner and I was forgotten about, right? Because of who I was. And the same thing happened to our people, when they had to hide, you know, when people were out looking for them.

Although some communities increasingly came to rely on formal employment, other, more remote communities continued to rely on hunting, trapping, fishing, and gathering berries and medicines. However, with Confederation, a program of road-building, centralization, and urbanization was begun. By 1975, two hundred outport communities had been abandoned, and 6 percent of the province's population had been relocated to urban centres (Royal Commission, 2003, p. 35). On the west coast, some communities were entirely overwhelmed by white incomers, and formerly remote Mi'kmaq communities were now accessible by roads. With increased roads came increased surveillance. Calvin White recalls what this process meant to his community at Flat Bay:

> Flat Bay was an isolated community all throughout and up until the 1960s. We did not have roads in the community until about 1956, we had electricity in 1960, and we had the phones in 1964. We did not feel any pressures of outside intervention into our communities prior to the late 1960s. In the mid-1960s because of the road and because of the availability of vehicles, and people to vehicles, [outside people were] transported into our communities. Our communities had been traditionally practicing

> a way of life. What I mean by that was the people lived off the land and the sea, so the hunting and fishing was not something that was regulated by an outside authority; it was something that was practiced and regulated by a community structure. After the road went through in the 1960s, we started having the controls from the police and wardens and all of these other things coming into our community. What we saw was the erosion of our way of life based on the European concept of conservation, which we argue to this day is as exclusive and as protective as what we ourselves were practicing in our communities. What happened was that prior to and during the time, the structure of community was one that various people in the community took responsibilities for. Midwives took responsibility for the births of the newborns, midwives also took responsibility to be there for the sick and dying, and then you have other people who took responsibility for the burial, and people who took responsibility for organizing the hunting parties . . . It was not until we were infiltrated by outside authorities who were inflicting their idea of governing onto us that it became apparent that our way of life was completely threatened. We realized there would have to be a structure to be able to challenge these authorities and that is when in 1968 and 1969 the band councils came into being and that is when the council and chief was formed . . . for the simple reason that the community needed structure to be able to deal with, oppose, and educate the outside interest that was now coming into our communities.

In many respects, during the 1960s, as white society penetrated those remaining traditional Mi'kmaq communities where people had maintained a subsistence economy, reliance on the land ended for those communities altogether. Communities that hunted together now had to rely on individual licences, communities that formerly had no refrigeration, so that a moose was divided up among people, now hunted for only their own family in season and froze the rest. The collective structures of communities began to erode – ironically, since these measures were supposedly in the interests of conservation.

For those individuals who resolutely continued to hunt, amendments introduced to the Wildlife Act in 1982 were drastic and unprecedented. Fines for first offenders increased to not less than $1,000 and not more than $5,000, with mandatory confiscation of the vehicle used in poaching. In default of payment, a jail term of one to six months would be imposed. A second-time offender, convicted within five years of the first offence, would receive a jail term of one to six months plus a fine of between $3,000 and $10,000 (McGrath, 1996, p. 81). The "war" on poaching, made at a time when caribou were on the increase and the moose population only slightly declining (p. 92), was followed by a significant decline in the numbers of domestic hunting permits issued and an increase in the numbers of out-of-province hunters. Overall, the changes in legislation and in licensing followed Newfoundland's efforts to expand outdoor tourism in the late 1970s, as part of an overall investment of $13 million in the tourism industry (p. 89). Throughout this process there have been inadequate studies of game (and fish) populations, and little attention paid to habitat destruction through clear-cutting; it is therefore unclear to what extent poaching is responsible for any demise in Newfoundland wildlife. However, the harshness of the legislation functions to effectively terrorize Mi'kmaq people into stopping their hunting and fishing.

Tony John, of Glenwood First Nation, has been the first Newfoundland Mi'kmaq to openly defy the Department of Fisheries and Oceans by asserting an Aboriginal right to fish for food and ceremonial purposes:

> We've challenged DFO five years ago. We said, "We have a right to take salmon for ceremonial and food purposes, and by god dammed we're gonna take 'em." And we went out, and put our net out and tried to take some salmon, and had a rough time with the community because our beef wasn't with the community, our beef was with the DFO for nonrecognition of our Aboriginal treaty rights. We got into some sore fights with the non-Aboriginal community committee and that was rightly so. They figured that we had no right, and we figured that we had a right, and you're going to get that head-butting anyway. But at the end of the day we were charged for illegally putting a salmon net into regulated waters. And they came over to us and were going to read us our rights, and we said, "That's your rights, our rights is right into my card. My band card says we have a right to those salmon; I'm going to stand by it." The council went to court for five years, the federal government knew that we were going to win this case, they reneged on it, they backed down. Brought back my net, and it's been sitting there ever since. So what does that tell me, as an Aboriginal person? I have an Aboriginal right to those salmon. And I will have those salmon. As in proving it and getting recognition for that thing, they don't want to deal with it, the same thing – the salmon is a federal issue. The feds didn't want to deal with the issue. So we did the same thing with the moose. I shot a moose and put a call to the wildlife to come in, call the RCMP. So then they called and said we're going to court tomorrow, we've got the date, the 25th of some month, around there. And I said fine, I'll see you there, no problem. Never did see a court summons and never did see a court date.
>
> You gotta look at the big picture . . . if Tony John's the only guy that's doing it, the rest of the Mi'kmaq doesn't go down there. So what more could you do? Without going and forcing the issue to the water or to the bush and saying, "Do whatever you said." But the thing is, the govern-

ment is reluctant to deal with it. But it's five, probably seven years now trying to get this Aboriginal right to fish for food, ceremonial purposes through the courts. And they're not moving.

In the interviews it was clear that some of the leadership, particularly on the west coast, believe that they will never be able to assert hunting and fishing rights, when much of the local land has been commercialized, and where their own community hunting grounds have for a long time been taken over by development or taken over by white licensed hunters. However, unequivocally, every individual who I interviewed emphasized that environmental protection was crucial – and indeed, that not only Mi'kmaq but other concerned community people had to begin to have a say in how the environment is dealt with.

This perspective also came out clearly in interviews concerning community empowerment generally. Individuals who spoke about encountering resistance from local whites when they asserted their rights were clear that the more empowered Mi'kmaq people are with federal recognition, the stronger the local economy would become, with their increasing role in it.

Mi'kmaq communities appear to be ready to wait for as long as they have to in order to have their rights as Aboriginal people recognized. Increasingly, their attention is turning to empowering the young to feel stronger about their Native identity – knowing that the children will have to take up the struggle when their Elders have passed on. The late Elder Nish Paul spoke about this:

Well, you know, the Mi'kmaq language is practically gone for us, right? You know – and I guess what they're trying today, to organize something, so we can send our younger generation out and have them learn the language, you know. It will never be the same. And I do a lot of that myself, you know, I'm a spiritual leader, and I go out and do the sweetgrass ceremonies and stuff like that, right? And it just seems like the younger generation, they're getting into that now, right.

An important part of this process is standing up to racist mythologies that have held Newfoundland Mi'kmaq in oppression for too long. Tony John speaks to the tremendous upsurge in community empowerment that takes place when people stand up to racism:

You know, run and hide, don't tell anybody who you are, and speak English, and dye your hair blonde and try to blend in, assimilate. And then we started learning that we were blamed for the demise of the Beothuks . . . then we put our hands over our face even more, and I said, "Well, geez, did

> my ancestors do that?" And then we said, "Well geez, well, that's what the white guy said, let's go see ourselves." And we reached in and can't find one shred of evidence, one shred saying that we done that. So what happened? It was fabricated. They blamed us. Why? Because they didn't want to blame their own selves. So we wore the bloody bonnet for a hundred years and our sisters and brothers and relatives were told, "Go hide because you killed off the Indians, you killed off the Beothuks." And it was a god dammed lie . . . "I'm Mi'kmaq man, no problem. There ain't nothing to be shy about, I've got nothing to hide, I'm not ashamed of nothing." And that grows, and that's grown like wildflowers today.

Conclusion

In listening to the voices of the leadership who were interviewed in the past couple of years, it seems clear that land is central not only to how Mi'kmaq people have known themselves but to the people's sense of their future. Whether the goal of different communities is environmental protection or asserting their Aboriginal rights to hunt and fish, it is clear that the land continues to be deeply connected with who the people are.

Meanwhile, pressures continue to be asserted on communities as myths and stereotypes about "Indianness" are employed by the federal government in an attempt to drive successive wedges through the nation, with some communities receiving federal recognition as being "Indian enough" to be registered under the Indian Act and get a reserve and others not. Failure to adequately establish "Indianness" results in groups of people being thrown back on the larger society to survive, where they face continued pressures to suppress culture and identity, making future recognition as "Indian" still more difficult to obtain. As the author's previous work (Lawrence, 2004) demonstrates, when formal legislation defines a group's identity as Indian, it has powerful repercussions on how Indianness is seen, in common-sense ways; Native peoples are required to conform to those common-sense stereotypes in terms of "purity" of lineage, retention of language, and reliance on the land for livelihood. If Newfoundland Mi'kmaq identity has involved a subsistence livelihood on the land, can a Mi'kmaq identity without land be sustained?

At a grassroots level, cultural renewal appears to be central to what many of the younger Mi'kmaq are attempting to engage with, through music, through relearning language, through reclaiming local spiritual sites, and through other actions, to actualize their identities as Newfoundland Mi'kmaq. All of these actions are necessary to building a strong sense of Mi'kmaq identity in the youth. The leadership are clear that the silencing that their parents faced, through racism, and the disconnection from the land that their generation faced cannot be allowed to go any further – the next generation needs to be strengthened, and relationship to land is a crucial part of this. Most of the people interviewed seem fully determined to wait out the government's stalling game – to continue to build a strong sense of Mi'kmaq identity in the youth, knowing that Newfoundland Mi'kmaq people are not going anywhere, and that their time will come.

Wel'alioq! Um sed nogumak.

(Thank you. All my relations.)

Notes

1 David McNab notes that both the Mi'kmaq of Cape Breton and of Newfoundland referred to the people of the "land across the water" (Ktaqamkuk) by that name. D. McNab, "The Mi'kmaq Nation of Newfoundland and Section 35 of the Charter of Rights and Freedoms," in R. Riewe and J. Oakes (Eds.), *Aboriginal Connections to Race, Environment and Traditions* (pp. 30-31), Winnipeg: Aboriginal Issues Press.

2 Personal communication with Gabe Marshall, Mi'kmaq Elder, Toronto, 1995. Marshall also drew a map to demonstrate how the rest of Mi'kmaki was divided into seven hunting territories, which included the islands in the Gulf of St. Lawrence, St. Pierre and Miquelon, and other islands, among them Prince Edward Island as Epekwidkt; and to demonstrate the division of present-day Nova Scotia into four territories: Piwktuk, Eskikewak'ik, Sipeknekatik, and Kespukwith; and the division of eastern New Brunswick into two territories: the south, called Sikniki, and the north (with the Gaspé region of Quebec) called Kespek, the last lands.

References

Alexander, D. 1980. Newfoundland's Traditional Economy and Development to 1934. In J. Hiller and P. Neary (Eds.), *Newfoundland in the Nineteenth and Twentieth Centuries: Essays in Interpretation* (pp. 17-39). Toronto: University of Toronto Press.

Anderson, K. 2000. *A Recognition of Being: Reconstructing Native Womanhood.* Toronto: Sumach Press.

Armstrong, J. 1998. Land Speaking. In S.J. Ortiz (Ed.), *Speaking for the Generations: Native Writers on Writing* (pp. 174-94). Tucson: University of Arizona Press.

Bartels, D. 1987. Ktaqamkuk Ilnui Saqimawoutie: Aboriginal Rights and the Myth of the Micmac Mercenaries in Newfoundland. In B.A. Cox (Ed.), *Native People, Native Lands: Canadian Indians, Inuit and Metis* (pp. 32-36). Ottawa: Carlton University Press.

Bartels, D., and A. Bartels. 2005. Mi'gmaq Lives: Aboriginal Identity in Newfoundland. In U. Lischke and D. McNab (Eds.), *Walking a Tightrope: Aboriginal People and Their Representations* (pp. 249-80). Waterloo, ON: Wilfrid Laurier University Press.

Benwah, J. 2005a. History of Mainland. http://www.jasenbenwah.ca/mainland.html.

–. 2005b. St. George's Bay Mi'kmaq/Nujio'qonik Mi'kmaq Communities. http://www.jasenbenwah.ca/index2.htm.

Calamai, P. 2005. Beothuk Mystery. *Toronto Star,* 6 August.

Clifford, J. 1988. *The Predicament of Culture: Twentieth Century Ethnography, Literature and Art.* Cambridge, MA: Harvard University Press.

Garroutte, E.M. 2003. *Real Indians: Identity and the Survival of Native America.* Berkeley: University of California Press.

Hanrahan, M. 2003. The Lasting Breach: The Omission of Aboriginal People from the Terms of Union between Newfoundland and Canada and Its Ongoing Impacts. In V. Young, E. Davis, and J. Igloliorte (Commissioners), *Royal Commission on Renewing and Strengthening Our Place in Canada: Collected Research Papers* (pp. 207-78). St. John's: Office of the Queen's Printer, Newfoundland and Labrador.

Henderson, James (Sakej) Youngblood. 1997. *The Mi'kmaw Concordat.* Halifax: Fernwood.

Jackson, D. 1993. *On the Country: The Micmac of Newfoundland.* (Gerald Penney, Ed.) St. John's: Harry Cuff.

Lawrence, B. 2004. *Real Indians and Others: Mixed-Blood Urban Native People and Indigenous Nationhood.* Vancouver: UBC Press.

Long, G. 1999. *Suspended State: Newfoundland before Canada.* St. John's: Breakwater.

Lyon, N. 1993. *The Mikmaqs of Newfoundland.* Ottawa: Human Rights Promotion Branch, Canadian Human Rights Commission.

Marshall, I. 1998. *A History and Ethnography of the Beothuk.* Montreal: McGill-Queen's University Press.

McGrath, D. 1996. Poaching in Newfoundland and Labrador: The Creation of an Issue. *Newfoundland Studies, 12*(2), 79-104.

–. 1994. Salted Caribou and Sportsmen Tourists: Conflicts over Wildlife Resources in Newfoundland at the Turn of the Twentieth Century. *Newfoundland Studies, 10*(2), 208-25.

Monet, D., and Skanu'u (A. Wilson). 1992. *Colonialism on Trial: Indigenous Land Rights and the Gitksan and Wet'suwe'en Sovereignty Case.* Philadelphia and Gabriola Island, BC: New Society.

Overton, J. 1980. Self-Help, Charity and Individual Responsibility. In J. Hiller and P. Neary (Eds.), *Newfoundland in the Nineteenth and Twentieth Centuries: Essays in Interpretation* (pp. 79-122). Toronto: University of Toronto Press.

Parker, J. 1950. *Newfoundland: 10th Province of Canada.* London: Lincolns-Prager.

Royal Commission (on Renewing and Strengthening Our Place in Canada). 2003. *Our Place in Canada.* St. John's: Office of the Queen's Printer, Newfoundland and Labrador.

Sheppard, B. 2006. Registration and Recognition for Mi'kmaq of Newfoundland. (Federation of Newfoundland Indians.) http://www.fni.nf.ca/news/Negoations%20Newletter.pdf.

Silko, L.M. 1996. *Yellow Woman and a Beauty of the Spirit: Essays on Native American Life Today.* New York: Touchstone.

Speck, F. 1982. *Beothuk and Micmac.* New York: AMS Press. (Orig. pub. 1922, Museum of the American Indian.)

Sturm, C. 2002. *Blood Politics: Race, Culture and Identity in the Cherokee Nation of Oklahoma.* Berkeley: University of California Press.

Walters, A.L. 1992. *Talking Indian: Reflections on Survival and Writing.* Ithaca, NY: Firebrand Books.

3

Why Is There No Environmental Justice in Toronto? Or Is There?

Roger Keil, Melissa Ollevier, and Erica Tsang

In 1906, Werner Sombart published his seminal essay on why there was no socialism in the United States. He compared the American case with that of Germany, where there was both an active labour movement and a strong social democratic party. He noted three main reasons, political, economic, and cultural, why the American society of the time had not developed what became a significant characteristic of European societies of the twentieth century. Sombart's set of explanations have been rightly criticized for their selectiveness, but his methodology is an intriguing precedent for a similar question we might ask today: Why is there no environmental justice (movement) in Toronto, or Canada for that matter? There are certainly political, economic, and cultural differences between the United States – often seen as the birthplace and major site of the environmental justice movement – and Canada, where no significant movement of this kind exists – at least under this heading (Debbané and Keil, 2004).

Canadian Challenges

Canadian and American researchers have made similar conclusions in their studies of environmental justice. According to Jerrett et al.,

> The similarity between the Canadian and US findings supports the idea that similar societal processes are in operation, with disadvantaged groups who live in low-cost housing being subjected to higher potential exposure . . . one exception to this similarity appears to be with the recent immigration variable. Although it is an imperfect measure of race, there are suggestions that environmental injustice in Canada may not have the same racial dimensions as it does in the US, the 1996 Census of Canada contains the first race variables, and appears to be an integral area for future research. (2001, p. 969)

But still, there are clear differences. The challenge for Canadian environmental justice researchers is to avoid simply borrowing American theoretical and methodological approaches that may be inappropriate to the Canadian context (Teelucksingh, 2001, p. 4; Debbané and Keil, 2004; Draper, 2001). Teelucksingh (2001) addresses the gap in the environmental justice literature by exploring the significance of race, class, and immigrant status as variables that contribute to people's different exposures to environmental risks in the City of Toronto (p. 3). By examining racialized spaces, she demonstrates the need to challenge the dominant thinking that race does not matter in Canadian contexts because most Canadian cities do not have American-style racial segregation. But racialization is latent and systematic in Canada (p. 5).

No matter how much we prefer to think of ourselves as a multicultural and diverse nation, racism and inequality do exist in this country. Similarly, no matter how much we prefer to think of ourselves as environmentalists, environmental problems do exist in Canadian cities, and disproportionately burden certain segments of the population. There is clearly a growing reality of socio-cultural segregation in Toronto, for example, which has led to tendencies of viewing multiculturalism as a force of separation rather than unity. In a fine aperçu, Toronto journalist Cameron Bailey has summarized this tendency: "In this city, we all go home to different suburbs. Kingston, Bangalore, Mogadishu. My feet are at home on the streets of Barbados. The looks I get there – and give – are warmer. Of course, my diaspora can be your nightmare" (Bailey, 2002). Toronto's view of itself as the most diverse city on the planet usually comes with the bravado of claiming normative superiority in questions of diversity, too. As Sheila Croucher (1997) has shown in a perceptive paper, Toronto multiculturalism has often been self-righteously compared with race relations south of the border. Multiculturalism as a policy has certainly had a mixed and contested history as the institutions of Canadian society have lined up to support what was originally conceived by the Trudeau government in 1971. But, as Wood and Gilbert (2005) contend, it may be more of an accidental discourse than planned scheme. More critical of the tradition of multiculturalism are Goonewardena and Kipfer (2005), who have pointed to the paper-thin veneer that multiculturalism spreads over racialized socio-spatial relationships in Toronto. The statistics do, in fact, bear out a growing mismatch between the higher normative goals of a just multiculturalism and the reality of social polarization. Grace-Edward Galabuzi has commented on this in a recent newspaper op-ed, which issues a stern warning that Canada needs to do "some bold thinking" around its future if it is to remain true to its stated goals of social justice and equity. Much of the danger to that goal stems from the increasing racialization of poverty: "In Toronto alone, the United Way reports that between 1980 and 2000, poverty among racialized groups grew by about 360 percent, compared with a decline of 28 per cent among the nonracialized population. The overall effect was to raise poverty in Toronto to 19 percent, well above the national average of 14 per cent" (Galabuzi, 2007).

Debbané and Keil have noted that "the term 'environmental justice' in the context of urban environments must be defined with this context for each site under study" and have argued against a "universalising use of the term" and instead for a local grounding of its use. They have also insisted on the recognition of the term's points of reference in a world constructed at various scales of meaning (e.g., policy environments from the local to the global). In comparing environmental justice work in the United States, South Africa, Namibia, and Canada, these authors have pointed out that it is possible to differentiate modes of racialization, spatialization, and political regulation in each national or urban context. It is important to keep this in mind for the Toronto case (Debbané and Keil, 2004; see also Kurtz, 2003).

Examining Environmental Justice Issues in Toronto

Toronto has suffered from numerous environmental problems over the years. In this section, we briefly examine some of those cases that have social justice implications. These areas include soil contamination, garbage and waste disposal, housing, industrial land use, and food and food security.[1] After examining the three communities of Parkdale, Mid-Scarborough, and South Riverdale in depth, Teelucksingh (2001) concluded that instances of environmental injustices and environmental racialization do exist, particularly in Mid-Scarborough and South Riverdale. Environmental risk was associated with the presence of industrial facilities (including emissions and impact zones) as well as health and safety concerns that disproportionately affect marginalized groups because of poorly maintained or dangerous housing (p. 299).

Soil Contamination

The community of South Riverdale has experienced numerous problems resulting from the contamination of its soil. This community was historically inhabited by poorer residents and, as a result, became the site of many of the city's industrial facilities. For instance, the Canada Metal Company had been a source of major environmental and health concerns for years, especially regarding lead poisoning. Members of the community grew frustrated by government inaction and eventually rallied themselves, arguing that the government would never have tolerated such living conditions in one of Toronto's wealthier neighbourhoods (Leboute, 1986). The general feeling was that because the community was poorer, industries were allowed to degrade the environment to an extent that would not have been accepted elsewhere.

Waste

Proposals for new waste sites, and expansions of existing ones, have often resulted in disputes in the communities where they are (or will be) located. Questions of fairness and equity are often central to conflicts over facility siting decisions (Fletcher, 2003, p. 67). The City of Toronto has a history of sending its garbage away because it has limited waste site capacity at local landfills. It ships waste long distances to communities that receive little or no economic benefit from it. Instead, these host communities are asked to bear the majority of environmental burdens associated with facilities such as landfills and incinerators; such cases raise spatial equity concerns. There are numerous examples of this, including proposals to ship waste to Kirkland Lake in northern Ontario and to Michigan. More recently, there has been concern over the Green Lane landfill in southwestern Ontario. Members of the Oneida Native Reserve have been vocal in their opposition to the placement of this landfill near their community. In each of these instances, the "people exhibited what many would regard as a classic NIMBY response, but their grievances also involved numerous fairness issues that are consistent with the issue of environmental justice" (p. 16). In the case of the Green Lane landfill, the fact that the landfill was to be located near a Native reserve was significant, and raised social equity concerns. Similarly, waste incinerators are an important source of air pollution and controversy. This was a big issue in the South Riverdale area of Toronto in the past. In fact, a group of citizens, Citizens for a Safe Environment, formed out of concern over the City of Toronto's proposed garbage burning incinerator in their community. Unfortunately, such social equity issues continue to plague poor communities because many heavily polluting facilities are purposefully located in these areas. A growing body of evidence supports the hypothesis that hazardous waste and other industrial pollutants disproportionately burden economically marginalized communities (pp. 11-12). Although the economic and other benefits of production, including employment and revenue generation, are distributed broadly across society, the burden of waste is not (p. 14). Thus, the potential exists for serious ramifications to be felt by the Toronto-area communities that experience waste-related environmental injustices.

Housing

There has also been concern about poorly maintained housing in the Toronto region. In particular, the community of South Parkdale has experienced insect infestations, inadequate garbage collection facilities, and housing that does not conform with fire and health standards. Similarly, residents of Mid-Scarborough suffered from poorly maintained government-subsidized housing, with cockroach and rat infestation, bad smells, broken windows, and poor ventilation. What makes this an environmental justice issue is the fact that these communities have a higher representation of immigrants, visible minorities, and incidents of low income (Teelucksingh, 2001, pp. 187, 213). Teelucksingh argues that environmental risk is associated with health and safety concerns arising from poorly maintained or dangerous housing that disproportionately affects marginalized groups in these communities (p. 299).

Industrial Land Use

There has been some apprehension concerning industrial land use in the Toronto region. High proportions of visible minority populations, recent immigrant populations, and low median income populations are located in close proximity to high-emission facilities in Toronto (Teelucksingh, 2001, pp. 161, 163). In fact, these populations occupy an inverse U-shaped area around the former City of Toronto. The U shape of potential environmental racialization includes areas of South Riverdale, East (North) York, and Mid-Scarborough, areas identified in research studies as sites of environmental injustices and high environmental risk (p. 163).

The Port Industrial area (of South Riverdale) is the only remaining heavy industrial zone within the City of Toronto. It has a long history of community activism in opposition to toxic pollution emissions. In fact, this led to the closing of some industrial facilities, diminishing the air, water, and soil pollution for those residents living nearest to the facilities. However, activism was dominated by white residents and the concern was mainly for these people, not for new immigrants or poorer residents. Ongoing soil contamination issues in this area have concerned often disadvantaged communities in the old industrial quarters of the city, while site-specific cleanup processes have replaced a more universal policy. In addition, "defensible science" has tended to replace previous concerns with public health in soil contamination conflicts (Desfor and Keil, 2004). Similarly, the Junction Triangle area of the city, bounded by Keele, St. Clair, and Dundas streets, has been plagued by environmental justice problems. This disadvantaged neighbourhood was the site of many high-polluting factories. In the past, plastics factories have dumped chemicals into the sewage system, leading to health problems in residents, including migraines, fatigue, dizziness, and vomiting. In addition, many complaints were made about the smell of chemicals, glue, and paint. Several community groups, including the Junction Anti-Pollution Group and the Bloor Junction Neighbourhood Coalition, were formed in response.[2] Trade union-initiated activism around the regeneration of industrial lands has also been connected to campaigns for green work and environmental production. This kind of activism tended to be working class and ethno-culturally diverse and differed considerably in terms of its sociological composition as well as its political direction from mainstream urban environmentalism (Keil, 1994).

Food and Food Security

Food and food-security issues are central to a broader understanding of environmental justice in Toronto. Represented most clearly by the work of the City's Food Policy Council under the directorship of Wayne Roberts, food production, consumption, and security issues have been (a) important to the overall sustainability of the city and (b) a major arena of real diversity as gardeners, urban farmers, and consumers, increasingly wary of corporate standardization and industrial contamination, have made food a showcase issue of survival in the global city. The complex physical and economic metabolisms of the food industry tie into existing patterns of class, race, gender, and immigration (see, for example, Keil and Boudreau, 2006, on issues of environmentalism and metabolism in Toronto).

As is evident from these examples, environmental justice issues do exist within the City of Toronto. Although not always recognized as such, this does not diminish the seriousness of the implications that arise for residents of these communities. It is important to understand the characteristics of Toronto that make it unique and therefore more vulnerable to instances of environmental injustice.

Poverty and Diversity in Toronto

Although Toronto has long been recognized as one of the most livable cities in the world because of its rich mixture of residents, safe streets, and community services (United Way, 2004, p. 1), its positive reputation has recently suffered: "Poverty is rising, and deepening, and the income disparity between rich and poor is widening. Toronto's population is growing much faster in the inner suburbs yet there has been no commensurate investment in social infrastructure" (p. 1). In particular, concern has been raised regarding the rapid and dramatic rise and intensification in the number of high-poverty neighbourhoods, particularly in North York and Scarborough (p. 2). Thus, not only has poverty increased but it has become more concentrated. Moreover, there has been a shift in the resident profile of higher poverty neighbourhoods, with poor visible minority and immigrant families making up larger percentages of the total poor-family population in these neighbourhoods than twenty years ago (p. 4). Furthermore, the presence of lone-parent families is also linked to injustice issues because this family group often constitutes a community of least resistance in urban studies; however, it has been given little attention in justice research. Future work on nontraditional groups should include a control for residential locations, with a manufacturing employment variable. There is little research as to whether lone-parent families are found disproportionately in high-exposure zones because their residential choice is influenced by the location of services for their particular family form (Buzzelli, Jerrett, Burnett, and Finklestein, 2003, p. 569). These trends are alarming for two reasons. First, the consequences of living in a poor neighbourhood are significant. Second, poor neighbourhoods can spiral into further decline, resulting in an increase in crimes and abandonment (United Way, 2004, p. 1). According to a 2003 Statistics Canada report, 43 percent of all new immigrants to Canada settled in Toronto. The majority of the poor now tend to be immigrants and visible minorities.

Ghettos in Canada?

We can no longer pretend that Canada does not suffer from discrimination, racialization, and social exclusion. Numerous studies indicate that there is a need to be concerned about the spatial concentration of poor minority groups in Canadian urban areas: "Poverty (or more accurately, low income) is not only growing in Canadian cities but is becoming increasingly concentrated in poor neighbourhoods. Not surprisingly, the spatial concentration of visible minorities, Aboriginals and recent immigrants is cited as one of a number of potential factors underpinning the growth of concentrated urban poverty" (Walks and Bourne, 2006, p. 274).[3] Moreover, researchers have stated that "growing income inequality . . . increases the odds of poor visible minorities . . . ending up in the lowest-cost, least-desirable neighbourhoods from which they cannot afford to escape" (p. 274). Toronto is often singled out as the city demonstrating the most obvious signs of such segregation. One explanation for this is that immigrants prefer to settle in large urban regions, and thus Toronto becomes home to the majority of new immigrants. For example, immigrants constitute 49.4 percent of Toronto's population (Good, 2005, p. 261). When discussing the existence of poor, urban, minority neighbourhoods, it is inevitable that the concept of the ghetto will be brought up. The definition of a ghetto, as applied in the United States, is that of "a residential district that both concentrates a particular racial or ethnic group and at the same time contains it, in that a majority of its members are forced to live there due to discrimination on behalf of the host community." In particular, there is often discrimination in the housing and labour markets (Walks and Bourne, 2006, p. 276). Much of the literature on ghettos comes from the United States, where racial segregation, particularly of the black and Hispanic population, is a major concern. Although stark racial segregation like that in the United States does not (yet) exist in Canada, there are indications of increasing inequalities in Canadian cities, especially Toronto. Whereas a high degree of racial concentration is not necessarily associated with greater neighbourhood poverty, Walks and Bourne did find that the four most segregated Canadian cities, which included Toronto, have higher levels of spatial polarization than the largest cities in Britain and Australia (pp. 285, 294). Furthermore, the authors found that "the concentration of apartment housing, of visible minorities in general, and of a high level of racial diversity in particular, do help in accounting for the neighbourhood patterning of low income" in Canadian urban areas (p. 273). Consistent with similar studies, the authors discovered that visible minorities are clearly concentrated in a small number of urban regions in Canada (p. 282). The authors were careful to differentiate between the segregated groups in Canadian and American cities. Unlike in the United States, there are no black or Hispanic polarized tracts in any Canadian CMA (census metropolitan area). Instead, South Asians and Chinese are the most segregated groups (p. 285). The authors argue that Toronto is the CMA with the largest visible minority population and the greatest proportion of its population in highly concentrated tracts. Moreover, it is the only city where segregation levels have increased in recent years (p. 295). In fact, the authors state that "if future trends indicate any movement towards increasing segregation and/or ghettoization, or alternatively further evidence of newer forms of ethnic enclaves . . . it would seem that Toronto would be the first place to look" (p. 295). David Hulchanski (2007) argues that Toronto's neighbourhoods are increasingly segregated "naturally" by market forces and government social policy on the basis of socio-economic status, skin colour, and housing tenure. Methodologically, he points to the need to differentiate between an outcast ghetto and an enclave. According to Peter Marcuse (1998), an outcast ghetto is a ghetto in which ethnicity is combined with class in a spatially concentrated area with residents who are excluded from the mainstream of the economic life of the surrounding society. Conversely, an enclave is a spatially concentrated area in which members of a particular population group congregate as a means of enhancing their economic, social, political, and/or cultural development (Hulchanski, 2007). Moreover, Hulchanski argues that we must undo the concentration of apartment housing, of visible minorities, and the growing income inequality and improve the least desirable neighbourhoods. Thus, while there is agreement that ghettos do not yet exist in Canada, there are indications that they are starting to form, and, thus, it is not too late to reverse the trend. One particularly persuasive strategy for addressing environmental justice issues is to link them to currently utilized strategies for improving cities.

Linking Smart Growth to Environmental Justice

The pursuit of metropolitan regional and neighbourhood equity is, in many ways, an extension of the movement for environmental justice. It seeks to address not only what communities are against but also what they are for: healthy neighbourhoods with convenient access to good schools, affordable housing, parks, and grocery stores; equitable public investments; and access to opportunity. This movement responds to two challenges that poor and marginalized communities and neighbourhoods face as they seek to improve their quality of life. The first is that the larger patterns of metropolitan development have undermined past neighbourhood-based efforts to remedy concentrated urban poverty, socio-economic issues, and racial isolation. The second challenge is to find systemic ways to link poverty alleviation to the larger, society-wide patterns of social, economic, and environmental development (Bullard, 2007, p. x). Advocates of regional and neighbourhood equity are influential within the smart growth movement; however, they have been critical of both environmental groups and the conventional development industry for their lack of attention and responsiveness to issues of race, class, and poverty. They recognize, though, that public debate about smart growth and the new metropolitan agenda provides a political context to build new allies in the effort to address the unmet needs of poor people, working people, and people of colour in ways that improve the quality of life for everyone (p. x).

Why should environmental justice advocates care about suburban sprawl, smart growth, or regional equity? One answer is in the obvious effects of sprawl itself, neatly summarized by Hank Savitch (2003): "We pay a steep price for the privilege of sprawl. In one way or another, exposure to smog has been implicated in a host of illnesses. These diseases tend to be chronic and range from asthma to bronchitis, and ultimately to the exacerbation of heart ailments, emphysema, and pneumonia. Ascertaining the cost of pollution to human life is enormously difficult, and the figures are widely discrepant. But we do know that those most hurt are children as well as racial or ethnic minorities in poorer neighborhoods" (p. 599). The smart growth movement is beginning to influence what happens in communities all across Canada. Many core smart growth principles focus on protecting the environment, using resources wisely, facilitating cooperation between cities and suburbs, and investing in and rebuilding our inner-city older suburbs (Bullard, 2007, p. 24). Therefore, environmental justice and progressive smart growth advocates share the same interests and concerns.

Smart growth and environmental justice have many compatible goals. Both movements are built on a set of principles that recognize the interconnection between humans and the environment. Both movements also recognize that where we live impacts the quality of our lives and our life opportunities. And the framers of both movements encourage community and stakeholder collaborations in planning and decision making. Both movements offer opportunities for building coalitions and alliances to break down artificial barriers in housing, employment, education, transportation, land use and zoning, health and safety, and urban investment (Bullard, 2007, p. 24).

The Toronto situation largely mirrors the general observations made by Bullard (2007) and others on the subject. One of the major issues in the Greater Toronto Area, or even the Greater Golden Horseshoe region, has been the rapid sprawl that has covered broad swaths of agricultural land while it creates huge socio-ecological inequities across the region (Desfor, Keil, Kipfer, and Wekerle, 2006). The Toronto region is developing increasingly stark socio-economic divisions that are linked to the joint processes of downtown gentrification and exurban sprawl. While the region's glamour zones in the exurban expanses toward the Oak Ridges Moraine and at the condominium-lined waterfront get all the attention in urban development discourse, a large city-in-between is showing signs of increasing distress.[4] Job growth in the City of Toronto was negative with 0.1 percent and wages fell by 0.7 percent between May 2006 and May 2007, while the unemployment rate sat at 7.2 percent. At the same time, the sprawling, low-tax, land-rich exurbs catapulted the entire CMA to a 2 percent growth rate and an increase of 1.7 percent in wages, as unemployment was lower overall, at 6.4 percent (Scoffield, 2007). At the surface, sprawl seems to pay, and those communities shouldered with a more traditional urban form and set of infrastructures seem to be losing out in intraregional competition. Poverty is concentrated in some denser neighbourhoods as sprawl seems to be rewarded by the actions of policy makers, investors, and consumers. Could smart growth be an antidote to this detrimental development? Toronto and Ontario are already incorporating smart growth principles in their official plans, but environmental and social inequality still exist within the city and the province and is likely to become more pronounced if no counteraction is taken. The environmental community is lauding the expected "mutually reinforcing" economic, social, and environmental benefits of the new initiatives for smart growth, but neither the plans put in place nor the environmental reports on their potential effect make a particularly strong connection with the overall discourse and practice of environmental justice (Winfield, 2006, p. 13). We can also speculate that the general tendency prominent in Canada to couch debates on social justice in an overall culturalist discourse of diversity (Goonewardena and Kipfer, 2005) clouds the racialized features of land-use change and urban development. In other contexts, explicit connections have been made between smart growth and environmental justice, as the empowerment of communities through the planning process as well as focused investment "on areas that need it most" are considered crucial to any smart growth initiatives (NGA Center for Best Practices, 2001, p. 1). In order to effectively implement smart growth principles and reverse environmental and social inequality, the City and Province must make greater efforts to link these two movements wherever possible. The current Greenbelt and Places to Grow plans as well as the Official Plan of the City have little to say about how to address these issues forcefully and directly, and the environmental movement in and around Toronto has not been forceful enough to make the connections.

Toronto: Missing Connections

Repeatedly, environmental justice concerns are framed solely as a narrowly defined single issue; for example, as an environmental, a health, an income, a race, or an immigrant problem (Teelucksingh, 2001, p. 44). In order to further environmental justice research in Toronto, related environmental and social problems must be framed as environmental justice concerns (p. 44). On a small scale, multi-issue and multiracial environmental justice organizations are beginning to emerge in Toronto with the specific mandate of striving for environmental justice. These organizations face the challenge that an environmental justice framework supportive of their efforts to mobilize around environmental social injustices, including the link between racialization and environmental problems, is not widely recognized in Toronto. Other grassroots organizations exist in Toronto that may not self-identify as environmental justice groups but which are creating and challenging social spaces in their efforts to secure healthy and affordable housing, safer and cleaner neighbourhoods, and open and accountable interactions with state bodies and corporations that have the power to affect the local environment (p. 62). In the past, there have been various efforts in Toronto and elsewhere in Canada to merge environmental and social struggles, but the success of these efforts has been spotty (Harter, 2004; Keil, 1994; Keil and Debbané, 2005). A prime example was the 2002 struggle to keep Toronto's drinking water in public hands, which brought together a broad coalition of labour, environmental, and social justice groups in a common campaign (Young and Keil, 2005). If the systemic disconnect of social and environmental concerns can partly be blamed on the state institutions that pit multiculturalism as a positive policy of diversity against universal and individualized concerns about poverty, movement groups in Canada have to accept part of the responsibility for this disconnect, too. The continuing whiteness of much of the urban environmental movement is a mirror image not just of the hegemony of whiteness in Canadian society but also of the (local) state institutions after which they fashion and scale themselves.

Conclusion: Linking Social Justice to Environmental Justice

So, why is there no environmental justice in Toronto, or is there? The response to this question has to remain ambivalent. Although there has been a general reluctance by academics, policy makers, and activists to accept that environmental injustices and racialization exist in Canada, the amount of research being conducted on this subject has increased in recent years. Specifically, there is rising evidence that such trends do, unfortunately, exist in Canada, especially in its larger urban areas, such as the Toronto region. As we have shown in this chapter, there are – substantially, if not by name – many environmental justice initiatives underway in Toronto.

We have concentrated on the intersections of urban and environmental issues here. Another approach may lead to more distinct responses regarding the nature of the politics that connect people in the social justice and environmental movements conceptually and through activism. Sherilyn MacGregor's research (2006) has shown, for instance, that women active in social justice work in Toronto have in fact been profoundly involved in environmental justice work, maybe without referring to themselves as environmentalists. The question, then, seems to be one of definition rather than substance. Under certain light, environmental justice activism comes into full relief, although the language used to name it does not comply.

Not forging links between issues of poverty, racialization, and environmental degradation makes it difficult to address problems and to reverse the situation. Smart growth and social justice issues should also be seen as an integrated principle of environmental justice and vice versa.

Increasing poverty levels and greater concentration of marginalized groups have environmental justice implications for the City of Toronto. For instance, industrial facilities are often located in areas where the poor and visible minorities are concentrated, resulting in health and environmental concerns for these residents. Unfortunately, the people inhabiting these communities do not possess the power, time, or resources needed to seek change. As a result, their concerns are not prioritized by municipal governments. Significant changes must be made in how the City addresses environmental justice issues if we wish to improve the plight of these marginalized groups.

For the City of Toronto, as a major policy actor in the urbanized area of Toronto, to thoroughly address environmental justice issues, we believe that the following actions and realizations are necessary. The City must recognize that inequality, racialization, and environmental injustices exist in Toronto. The perception that Canada is a multicultural society has led to the common belief that racism is a problem only south of the border. The persistence of this myth has hidden the processes of racialization that exist in Canada (Debbané and Keil, 2004, p. 218). Therefore, municipal programs must be designed to acknowledge and address the interconnections of class, race, and space. Issues of poverty, health, and the environment in particular have potential to be linked. For example, community gardens could be developed, which would address these three issues. Youth in poor communities could be taught how to grow their own food in vacant lots. This would also allow them to fulfill volunteer requirements while fostering community engagement and activism. To prevent the formation of ghettos within the City of Toronto, we need to undo the concentration of apartment housing and visible minorities, address the growing income inequality, and improve the least desirable neighbourhoods within the city (Hulchanski, 2007). This would be the most effective way to address both environmental and social injustices (United Way, 2004, p. 14). There is a need for the City to contact relevant community groups and form relationships, or, more specifically, partnerships, with them. There are already many community groups working at the local level that address environmental justice issues but which have not yet been labelled as environmental justice organizations. There is a wealth of knowledge and expertise at the local level that can be drawn upon. A web of contacts has already been established and it should be utilized by the government to attain maximum potential in reversing current trends: "A diversity of groups from civil society must be allowed to bring their understanding, language, and problem definition to the struggle" of addressing social and environmental problems (Desfor and Keil, 2004, p. 168). Thus, the state must increase the power and responsibilities accorded to these groups. Problem definition should not be limited to technical discussions among experts but must be open to civic groups, and affirmative actions must be taken to support groups with fewer resources (p. 169). Not only must the government include these groups in discussions prior to the creation of new city policies but it must also support weaker groups and provide them with the means necessary for coming to the table. As well, it must make efforts to include as many stakeholders as possible. As Desfor and Keil (2004) note, "By including a variety of groups in a political process, the fractured, complex, and frequently contradictory character of environmental problems may be exposed. Debate among a broad range of interests may be able to focus on the desirability of social actions. In such a way, a process of identifying preferred social arrangements may be able to determine which actions should be taken" (p. 169). One practical step toward this would be updating the inventory of Toronto's environmental groups to include a breakdown of groups by level of jurisdiction (local, provincial, federal, international), as well as whether government or community-formed and -run and to identify those groups that specifically concern themselves with environmental justice.

By acknowledging the central role of citizens as the ultimate providers and consumers of policies, and by ensuring that citizens are well informed and aware of their civic responsibilities, a balance may be struck between the costs and benefits of economic development and growth, environmental protection and social equity (Agyeman, Bullard, Evans, 2003, p. 306). There is a need to frame environmental rights as a significant component of human rights. Following Alain Lipietz, Agyeman, Bullard, and Evans (2003) note that new rights and obligations should be incorporated within social norms (p. 10). This would involve the recognition of new rights, new bearers of rights, and new objects of rights (p. 10). The City of Toronto can strengthen human rights, as well as environmental justice, by recognizing the principles inherent in this kind of thinking. Emerging co-activism on sustainability and environmental justice issues can be found in local fights for just transportation, community food security, and sustainable communities and cities. In Toronto there is a growing number of examples of cooperative endeavours at the local level between environmental and social justice organizations (Keil, 1994; Desfor and Keil, 2004; Young and Keil, 2005).

The mayor of Toronto, David Miller, has started a campaign to make Toronto the greenest city in North America (Gorrie, 2007). A series of technical and governance advances are intended to provide the conditions for an increasingly carbon-neutral municipality. Part of the ambitious program's plan was to create an environmental roundtable, designed to advise the city council and the bureaucracy on environmental matters. The study that underlies this chapter is one of the outcomes of this process. Toronto also has an activist city council, many members of which – such as the suburban conservationist Glenn De Baeremaeker and the downtown progressive Gord Perks – were elected on explicitly environmentalist agendas. In addition, the City boasts a spectrum of the most powerful and innovative environmental organizations in the country, including the Toronto Environmental Alliance. Still, many discussions that have sprung from the initiatives fronted by these actors continue to view issues of ecological modernization as only a series of technocratic and market reforms. The structural social inequities in ecological conditions are rarely addressed. Among them are the lasting and systemic discrepancies in housing stock and tenure; although roughly half of Torontonians live in owner-occupied single-family homes and the other half in rental high-rise apartment towers or condos, most environmental innovations are geared toward the former only. Also, there is a growing sense of transit injustice, as the transit system is geared toward the wealthy residential areas and the downtown core, which leaves the less privileged populations in the inner suburbs in a lurch (a systemic problem to be addressed in the city's recent TransitCity proposals). Further, the dramatic pace of gentrification that has the city in its grip creates urban environments that are not just polarized and segmented today but will be on diverging trajectories in the future. The Official Plan, which lays out this future, is often seen as the blueprint for a city that privileges capital over communities, the exchange value of the city over its use value (Blackwell and Goonewardena, 2003). Ultimately, citizens are the main beneficiaries of sustainable development – economic development and growth, environmental protection and social equity – therefore, it is natural that citizens should be consulted and play an active role, if not the principal role, in the decision-making process on matters regarding these three main components of their living environment (Agyeman, et al., 2003, p. 306). Thus, in order for the City of Toronto to provide maximum protection against environmental injustices, it must increase the power and responsibilities of inhabitants in shaping official strategies, as well as implementation and follow-through.

Notes

We acknowledge funding from the City of Toronto and the Faculty of Environmental Studies for this research. A more detailed report on environmental justice in Toronto that was produced by the City Institute at York University in April 2007 can be found at www .yorku.ca/city.

1 For a more comprehensive examination of some of these issues within the context of three specific Toronto-area communities (Parkdale, Mid-Scarborough, and South Riverdale), see Cheryl Teelucksingh's study (2001) on environmental injustices, *In Somebody's Backyard: Racialized Space and Environmental Justice in Toronto.*

2 These issues were documented in a 1996 video directed by Deedee Slye (1996), *Tales from the Triangle.*

3 For a useful definition of the "wooly and shifting notion" of the ghetto, see Loic Wacquant's entry in the *International Encyclopedia of the Social and Behavioural Sciences,* at http://sociology.berkeley.edu/faculty/wacquant/wacquant_pdf/GHETTO-IESBS.pdf.

4 The term *glamour zone* is borrowed from Douglas Young.

References

Agyeman, J., R. Bullard, and B. Evans. 2003. *Just Sustainabilities: Development in an Unequal World.* Cambridge, MA: MIT Press.

Bailey, C. 2002, 4 July. The Masks We Wear: For All Our Complexity; T.O. a Cultural Ghetto. *Now Magazine,* 19.

Blackwell, A., and K. Goonewardena. 2003. The Poverty of Planning: Tent City, City Hall and Toronto's New Official Plan. In R. Paloscia (Ed.), *The Contested Metropolis* (pp. 221-24). Basel: Birkhäuser.

Bullard, R. (Ed.). 2007. *Growing Smarter: Achieving Livable Communities, Environmental Justice, and Regional Equity.* Cambridge, MA: MIT Press.

Buzzelli, M., M. Jerrett, R. Burnett, and N. Finklestein. 2003. Spatiotemporal Perspectives on Air Pollution and Environmental Justice in Hamilton, Canada, 1985-1996. *Annals of the Association of American Geographers, 93*(3), 557-73.

Croucher, S. 1997. Constructing the Image of Ethnic Harmony in Toronto, Canada: The Politics of Problem Definition and Nondefinition. *Urban Affairs Review, 32*(3), 319-47.

Debbané, A.-M., and R. Keil. 2004. Multiple Disconnections: Environmental Justice and Urban Water in Canada and South Africa. *Space and Polity, 8*(2), 209-25.

Desfor, G., and R. Keil. 2004. *Nature and the City: Making Environmental Policy in Toronto and Los Angeles.* Tucson: University of Arizona Press.

Desfor, G., R. Keil, S. Kipfer, and G. Wekerle. 2006. From Surf to Turf: No Limits to Growth in Toronto? *Studies in Political Economy, 77,* 131-56.

Draper, D. 2001. Environmental Justice Considerations in Canada. *Canadian Geographer, 45*(1), 93.

Fletcher, T.H. 2003. *From Love Canal to Environmental Justice: The Politics of Hazardous Waste on the Canada-US Border.* Peterborough, ON: Broadview Press.

Galabuzi, G.-E. 2007, 8 July. Toward a Decade of Hope and Prosperity. *The Toronto Star.* http://www.thestar.com/article/233299.

Good, K. 2005. Patterns of Politics in Canada's Immigrant/Receiving Cities and Suburbs; How Immigrant Settlement Patterns Shape the Municipal Role in Multiculturalism Policy. *Policy Studies, 26*(3/4), 261-89.

Goonewardena, K., and S. Kipfer. 2005. Spaces of Difference: Reflections from Toronto on Multiculturalism, Bourgeois Urbanism and the Possibility of Radical Urban Politics. *International Journal of Urban and Regional Research, 29*(3), 670-78.

Gorrie, P. 2007, 17 February. Toronto's Green Blueprint. *Toronto Star.*

Harter, J.H. 2004. Environmental Justice for Whom? Class, New Social Movements, and the Environment: A Case Study of Greenpeace Canada, 1971-2000. *Labour, 54,* 83-119.

Hulchanski, J.D. 2007. Ghettos of the Rich and the Poor: Is This Where Toronto Is Headed? In *Mapping Neighbourhood Change in Toronto.* Toronto: Centre for Urban and Community Studies, University of Toronto. http://www.urbancentre.utoronto.ca/gtuo/.

Jerrett, M., R.T. Burnett, P. Kanaroglou, J. Eyles, N. Finkelstein, C. Giovis, and J.R. Brook. 2001. A GIS – Environmental Justice Analysis of Particulate Air Pollution in Hamilton, Canada. *Environment and Planning A, 33*(6), 955-73.

Keil, R. 1994. Green Work Alliances: The Political Economy of Social Ecology. *Studies in Political Economy, 44*, 7-38.

Keil, R., and J. Boudreau. 2006. Metropolitics and Metabolics: Rolling Out Environmentalism in Toronto. In N. Heynen, M. Kaika, and E. Swyngedouw (Eds.), *In the Nature of Cities: Urban Political Ecology and the Politics of Urban Metabolism* (pp. 41-62). New York: Routledge.

Keil, R., and A.-M. Debbané. 2005. Scaling Discourse Analysis: Experiences from Hermanus, South African and Walvis Bay, Namibia. *Journal of Environmental Policy and Planning, 7*(3), 257-76.

Kurtz, H.E. 2003. Scales Frames and Counter-Scale Frames: Constructing the Problem of Environmental Injustice. *Political Geography, 22*(8), 887-916.

Leboute, R. 1986. On the Brink. *New Internationalist, 157.* http://www.newint.org/issue157/brink.htm.

MacGregor, S. 2006. *Beyond Mothering Earth: Ecological Citizenship and the Politics of Care.* Vancouver: UBC Press.

Marcuse, P. 1998. Sustainability Is Not Enough. *Environment and Urbanization, 10*(2), 103-11.

NGA Center for Best Practices. 2001. How Smart Growth Can Address Environmental Justice Issues. http://www.nga.org/portal/site/nga/menuitem.9123e83a1f6786440ddcbeeb501010a0/?vgnextoid=7f6b5aa265b32010VgnVCM1000001a01010aRCRD.

Savitch, H. 2003. How Suburban Sprawl Shapes Human Well-Being. *Journal of Urban Health: Bulletin of the New York Academy of Medicine, 80*(4), 590-607.

Scoffield, H. 2007, 10 July. Home Tax Called "A Recipe for a Slowing Economy." *Globe and Mail*, p. A13.

Sombart, W. 1906. *Warum gibt es in den Vereinigten Staaten keinen Sozialismus?* Tübingen: Mohr.

Teelucksingh, C. 2001. *In Somebody's Backyard: Racialized Space and Environmental Justice in Toronto* (PhD dissertation, University of Toronto).

United Way of Greater Toronto and the Canadian Council on Social Development. 2004. *Poverty by Postal Code: The Geography of Neighbourhood Poverty 1981-2001* (Executive summary). Toronto: United Way of Greater Toronto and the Canadian Council on Social Development.

Walks, R.A., and L.S. Bourne. 2006. Ghettos in Canada's Cities? Racial Segregation, Ethnic Enclaves and Poverty Concentration in Canadian Urban Areas. *Canadian Geographer, 50*(3), 273-97.

Winfield, M. 2006. *Building Sustainable Urban Communities in Ontario: A Provincial Progress Report.* Toronto: Pembina Institute.

Wood, P., and L. Gilbert. 2005. Multiculturalism in Canada: Accidental Discourse, Alternative Vision, Urban Practice. *International Journal of Urban and Regional Research, 29*(3), 679-91.

Young, D., and R. Keil. 2005. Urinetown or Morainetown: Debates on the Reregulation of the Urban Water Regime in Toronto. *Capitalism, Nature, Socialism, 16*(2), 61-84.

4
Invisible Sisters: Women and Environmental Justice in Canada

Barbara Rahder

Women are the backbone of the environmental justice (EJ) movement in the United States according to Peggy Shepard of West Harlem Environmental Action, Inc., and chair of the second Environmental Justice Summit held in Washington, DC, in 2002.[1] Indeed, women make up the majority of EJ activists in the United States, where they are the mainstay in both leadership roles and in local grassroots organizations (Adamson, 2002; Stein, 2004). This makes sense given that women – particularly women of colour – are the most likely to be affected by poverty and, consequently, the most likely to be exposed to the environmental hazards, risks, and other systemic disadvantages associated with poverty and space in the United States. These same environmental burdens hit Canadian women of colour hardest, too, though you might not know it.

Canadians share similar political-economic and ideological terrain with the United States, but our histories, our demographics, and the social and spatial processes guiding development of our territories are quite different. These differences, particularly the dramatic differences in patterns of racial segregation, have tended to obscure EJ issues in Canada to some extent, including the roles of women in EJ struggles. But these social and spatial patterns are changing in Canada. As Canadian cities become more diverse, we are witnessing increasing ethno-racial segregation and a well-documented decline in the life circumstances faced by low-income ethno-racial communities, and particularly women. While EJ issues have always been here, they are now becoming more visible.

The literature on EJ rarely examines gender despite that much of it mentions the trinity of "race, class, and gender." It typically acknowledges the importance of women's roles in the movement in the United States, and then concentrates only on the first two of these dimensions.[2] A feminist analysis brings the challenges faced by women to the fore, particularly the challenges faced by women marginalized and excluded from public policy making. By examining EJ issues within the context of a global political economy and the historically specific lived experience of local communities, we are better able to see how and why women are so crucial to any movement for environmental and social change.

Although Canadian patterns of environmental injustice have been well hidden, this is changing. What will happen next? Will we face a "dark age ahead," as predicted by Jane Jacobs (2004), or will we rejuvenate the environmental justice movement in Canada, with women of colour at the fore? A convergence of events and potential strategies make this an opportune time to look again at our options.

The Global Political-Economic Terrain

Environmental justice is not simply about the systemically racist distribution of toxic waste in communities of colour, though this is the origin of the movement and remains highly relevant in Canada, as in the United States. Environmental justice is about the structural and spatial inequities of production and reproduction in a neo-liberal political economy. All of the communities caught in the systemic trap of social and spatial marginalization, whether from race/ethnicity, class, gender, sexuality, ability, or some other form of oppression, are vulnerable. A feminist analysis of environmental justice must look beyond the experiences of women as a homogeneous group and examine the disparate lives of women systematically excluded from participation in public life. If we examine the spatial distribution of environmental benefits, as well as the distribution of poverty, health risks, and environmental contamination, there can be little doubt that there are structural problems in the underlying system of distribution.

Marginalized peoples and impoverished countries have not created the large-scale environmental problems plaguing the planet, though they must cope with the brunt of these problems. Environmental injustices are literally the *buy-product* of the wealthiest and most powerful peoples and countries of the world. The challenge in Canada involves our unwillingness to acknowledge the extent of environmental injustice within our own cities and country, our inability to come to grips with the need for a dramatic redistribution of wealth and resources – both within Canada and globally – and our reluctance to admit that women marginalized by poverty, racism, violence, and a host of other experiences systematically excludes them from participation in relevant decision making.

In order to promote EJ, we must critically reflect on the movement's successes and failures, and clearly situate environmental problems within the context of the global capitalist political economy. Although many of the most successful EJ efforts in the United States, for example, have been lead by women whose homes, families, and communities were threatened, their success at displacing toxic industries and waste is now an increasing threat to people of the global South (Pellow and Brulle, 2005). If we don't want EJ efforts to simply displace problems onto more vulnerable populations, with fewer environmental protections, then structural political-economic change and broader alliances are required.

McLaren (2003), a major proponent of a conceptual tool to address global issues of inequitable distribution called "environmental space," argues that overconsumption and waste by the wealthy, along with overexploitation by multinational corporations, are problems that cannot be solved or paid for by the poor, yet the process must include them and be both democratic and just.[3] Although the concept of the environmental footprint tells us how much of the earth's surface we use to support our current way of life (and North America's footprint far exceeds that of anywhere else in the world),[4] the concept of environmental space attempts to set minimum and maximum ranges of sustainable resource consumption, for example, on potable water use. Agyeman (2005) and Buhrs (2004, cited in Agyeman 2005) add an important proviso that contextualizes this concept within the struggle for social and economic justice. Agyeman further suggests adding similar floors and ceilings to social and economic entitlements, defined as *community space,* as well as limits to unsustainable consumption levels.

McLaren (2003) goes on to assert that we could achieve just sustainability if we work collectively to meet these global limits to growth by 2050, adding as an afterthought that equality for women may be the key to reducing global inequity and overconsumption. He notes that studies have shown over and over that as women's education and income improve, so too does their ability to control birth and, consequently, their ability to adequately care for their children, their communities, and their environment. Instead of an afterthought or an aside, then, justice and equity for women must be a basic tenet of any further conceptualization of environmental space or community space.

But some notions of sustainability are incompatible with social justice, while others depend upon it. Clearly, to understand what is meant by "just sustainability," we need to distinguish between the different values that underlie various notions of sustainability and of social justice. Dobson (2003) identifies three key types of each. He categorizes the sustainability literature as focused on sustaining critical natural capital, sustaining biodiversity, or sustaining the value of natural objects. At one end of the continuum, people value only that which sustains human life, while at the other end, people inherently value all natural objects. In a similar vein, Dobson categorizes debates in the social justice literature as involving distribution according to need, distribution according to merit, or distribution according to entitlement. The underlying values here are again fundamentally different – ranging from communitarian to individualistic – and represent conflicting ways of understanding social justice and of deciding upon what is fair.

Gunder (2006) argues that as long as the notion of sustainability is used to cover up for economic growth as usual, it will be a pernicious use of the term. I don't disagree with this point, but I think Gunder, like many others, assumes that we must choose between environmental sustainability and alleviating poverty, at least in part because the wealth and privilege of the so-called developed countries is seen as unassailable. But if the problem is primarily one of inequitable distribution, then it is the overconsumption, spectacular waste of resources, and obscene profits of multinational corporations that we must confront. In other words, it is the highly problematic "entitlement" version of social justice, noted by Dobson, that preserves the privileges of Western culture and is the major stumbling block to sustainability. *Just* sustainability depends on a communitarian, rather than an individualistic, notion of social justice.[5]

Agyeman, Bullard, and Evans (2003) note that the problems are much clearer than the solutions. They argue that we need to challenge the dominant paradigm that allows global polluters to make mega-profits while the costs are borne most heavily by women and people of colour. They suggest that issues of sustainability must be merged with the environmental justice movement such that human rights are reconceived as collective and substantive rights to a clean environment and procedural rights to participate in decision making about the environment.

As long as women are excluded, undervalued, disregarded, ignored, or otherwise made invisible, sustainability and EJ will be impossible.

The Lines Are Not Always Clear

Key differences between Canada and the United States have tended to obscure EJ issues in Canada. The United States has a long history of institutionalized racism, predatory planning, and racial segregation that gave rise to a vibrant civil rights movement in the 1960s and an equally dynamic, if arguably less successful, environmental justice movement ongoing since the 1980s.[6] In Canada, the history of racism has been better masked and the patterns of ethno-racial segregation less marked, at least outside the federal system of reserves for Native peoples and the former African-Canadian community of Africville, Nova Scotia. Changes in immigration over the past twenty years or so, however, have also changed urban residential patterns and experiences.

Native peoples in Canada have been faced with EJ issues from the time Europeans first landed in North America more than six hundred years ago. They have used the language of EJ at least since the 1990s (D. Jacobs, 1992; Richardson, Sherman, and Gismondi, 1993). There have been numerous cases of Native communities being exposed to extremely hazardous substances in relatively remote parts of the country, including dioxins from paper mills and mercury in fish. None has been starker than the case of contaminated breast milk among Inuit women in the Canadian Arctic, which has been linked to contaminants produced by Midwestern US factories. Carried high on air currents for hundreds of kilometres, these contaminants are precipitated into the air and water in the Arctic. Yet Inuit women have no legal recourse for the environmental and health problems that they and their children will face for many years to come (Lucas, 2004). A feminist analysis, argues Lucas, would "envision a fuller remedy" and make more equitable outcomes feasible (p. 202).

The story of Africville is unique in Canadian history.[7] In the early nineteenth century, it was a small town outside Halifax founded and occupied exclusively by the descendants of former slaves. As Halifax grew, Africville was annexed into the city but never provided with a sewage system, garbage pickup, streetlights, or other amenities normally distributed by the municipality. The City located slaughterhouses and the municipal dump in Africville and dug a ditch to drain sewage from the city's communicable disease hospital directly through the neighbourhood, but never managed to pave its streets or supply its residents with running water. This was the only exclusively black neighbourhood in any Canadian city and, perhaps not surprisingly, suffered a fate similar to black ghettoes in the United States during the urban renewal craze of the 1960s. That is to say, by 1970, the entire community had been displaced and dispersed, their homes and churches torn down, to make way for redevelopment that would benefit only the white majority.

While other low-income neighbourhoods in cities across Canada had similar, if not nearly so dire, circumstances thrust upon them, none were exclusively, or even predominately, black neighbourhoods. In Toronto, the low-income neighbourhoods of South Riverdale and the Junction Triangle were traditionally home to predominately white, working-class families. These families faced exposure to high levels of toxic emissions from paint, chemical, and other noxious factories, and both communities have organized and fought, at various times, to address these issues and the resulting threats to community, and particularly children's, health.

The lack of clearly delineated racial segregation in Canadian cities doesn't mean that environmental justice isn't an issue here. It just hasn't been as readily apparent as in the United States. On the whole, segregation in Canada cities has been far more clearly delineated along lines of income or class than by race or ethnicity, though this is changing as the ethnic diversity of cities changes. For example, only about one-third of all Native peoples live on reserves in Canada, while almost half currently live in cities.[8] Apart from Africville, there have been no predominately black urban neighbourhoods, so far, though the vast majority of black Canadians live in cities.[9]

A number of recent studies have shown that income disparity is increasing in Canada and that this disparity hits ethno-racial minorities hardest (Ornstein, 2000; United Way, 2004; Inside the US Income Gap, 2007). According to *If Low Income Women of Colour Counted in Toronto,* a report for the Community Social Planning Council of Toronto, "poverty rates among Toronto's communities of colour are double and triple those faced by peoples of European descent. For racialized women, the situation is alarmingly severe. In a number of communities, more than three-quarters of sole-support mothers and single women are poor" (Khosla, 2003, p. 8). The pattern of poverty is also becoming more spatially concentrated, particularly in Toronto, where racial/ethnic minorities and recent immigrants are increasingly segregated (United Way, 2004). Although Canadian cities have no large-scale black or Hispanic ghettoes, as in the United States, we do have increasingly segregated areas of South Asian and Chinese Canadians (Fong and Shibuya, 2000; Walks and Bourne, 2006), as well as numerous low-income neighbourhoods with remarkably mixed ethnicities.

For instance, as Teelucksingh (2002) makes clear, Parkdale is a Toronto neighbourhood with a mix of incomes and ethnicities, and an excellent example of "the dynamic spatiality of environmental justice problems" (p. 13). She notes that increasing proportions of immigrants, of visible minority groups, and of low-income people are mixed in with long-time residents and young professionals. She points out that industrial activities, urban development, and their associated environmental risks have also shifted and changed over time. Her point is that marginalized people and environmental problems do not necessarily share the "fixed spatial configurations" described in literature on the EJ movement in the United States (p. 2). The relationship between social and spatial structures has always been just as real and just as important in Canada, if more muted. But the racialized lines and threads of this relationship are just beginning to show their place more sharply in the urban landscape, literally.

Khosla (2003) notes that, women are waging quiet struggles to survive in these circumstances. They face "staggering" loneliness, isolation, and often have no knowledge of their most basic rights (pp. 51-53). Her study observes that there are few accessible, secular places for women to gather for healthy discussion and socializing – community space is often costly or reserved for recreation programs. The costs of housing, transit, and daily living keep low-income women isolated from one another in their own homes. Racialized low-income women of colour, particularly immigrant women, often face worse experiences, including being poorly treated by social workers, police, and others (p. 12). This is clearly an issue of environmental injustice.

So, why haven't Canadian women organized an EJ movement like in the United States? The United Way (2004) has documented how low-income women, even those who are employed, are slipping further and further behind. The *Toronto Star* reports that children (and presumably their mothers) are the largest group of food-bank users in Ontario (Growth of Food Banks Signals Rise in Poverty, 2006). Khosla (2003) reports single mothers and immigrant women are the most "stuck" in their homes, often dependent on their children for their interactions with the outside world, and lacking time, resources, skills, and social supports. It is not surprising then, that these women report feeling hopeless. Khosla sums up this situation well by saying that women are overworked, underpaid, and too stressed or swamped to get involved in advocacy (pp. 13-23). While low-income women of colour want to be full participants, they have few opportunities or spaces where they can get together in order to press for needed change (p. 69). Their isolation from one another and from other cultures and communities means that they are virtually invisible when it comes to public policy making (p. 33).

Invisible Women

Canadian women of various classes and races are actively engaged in addressing environmental issues, as well as broader issues of social justice and quality of life, though their organizing and activism is rarely recognized as part of an environmental justice movement. In fact, MacGregor (2006) found that the majority of women activists she interviewed in Ontario said that women activists outnumber men, just as they do in the United States. But, if women are the major participants, why isn't this recognized? Given that women make up 90 percent of active members and 60 percent of leadership in the EJ movement in the United States (Stein 2004, p. 2), why aren't women more widely acknowledged?

Several people have speculated about why women are so invisible in the EJ literature. Unger (2004) suggests that limited attention has been paid to women in the EJ literature, despite that they have been a persistent force throughout history, because much of the work they have done has not been recognized as EJ work or as the precursors to present day EJ activists. Kaalund (2004) illustrates the historic role played by black women, even when no one was paying attention, and argues that women haven't been highlighted in the movement for strategic purposes, in order to build an inclusive movement.

Although there is no doubt that women's contributions have been overlooked, in part out of neglect or oversight, and possibly for strategic reasons (however misguided), there is a more compelling reason put forth by Simpson (2002). She notes that there are many women of colour in leadership positions in the EJ movement who are neglected in the literature and by the press because they lack traditionally recognized credentials. Often low-income women become involved in environmental justice issues because of personal loss or threat. Many women are reticent to speak out publicly, and those that do face an uphill battle, particularly if they are poor, uneducated, and black. Simpson recounts the story of Doris Bradshaw and her work on women's reproductive cancers. She describes the systematic ways in which the claims of low-income women are dismissed as special interests, or worse, the way women such as Doris are demonized by the press as ignorant and self-serving. MacGregor's interviews (2006) with women activists in Ontario similarly found stereotyping a common means of discrediting women, particularly homemakers, working-class women, and women of colour. Low-income women of colour continue to face the most appalling bias – a bias well hidden by a continued and socially constructed pretence toward such things as credibility, neutrality, and authority.

How are these standards of credibility and authority established and reified? How is it that our beliefs and attitudes become so colonized that we don't see through the smog and subterfuge so often perpetrated by the mainstream media? The social mechanisms that simultaneously oppress women and racialized minorities are so engrained that they've become second nature. Gaard (2004, pp. 22-24) builds on the work of Sedgwick and Plumwood,[10] outlining some of the specific techniques used to reinforce these largely unconscious systems of oppression. I've paraphrased and modified some of Gaard's wording and added examples to illustrate how unconscious these techniques may become:

(1) *Backgrounding:* when the dominant group relies on the services of the other but simultaneously denies this dependency, such as when the dominant group assumes that "women's work" in the home isn't work at all and has no value, economic or otherwise;
(2) *Radical exclusion:* when the dominant group magnifies the differences and minimizes shared qualities with the other, such as when the dominant group portrays people of colour as potential criminals or terrorists;
(3) *Incorporation:* when the dominant group's qualities are taken as the standard and the other is defined in terms of her possession or lack of those standards, such as when the dominant group claims that "rationality" is superior to all other ways of knowing;
(4) *Instrumentalism:* when the other is constructed as having no ends of her own, as if her sole purpose is to serve as a resource for the dominant group, such as when the dominant group presents mothers' motivation as exclusively focused on the care of children and others; and
(5) *Homogenization:* when the dominated class of others is perceived as uniformly homogenous, such as when the dominant group depicts low-income people as stupid or lazy.

These techniques systematically undermine the credibility of women, particularly low-income and/or racialized women, whether consciously or unconsciously. Oppressive ideas and ideologies are as toxic to communities as any physically contaminated environment. By illuminating the relationship between violence done to women, to peoples of colour, to low-income communities, and to land, we are sometimes able to inspire women – and men – to get involved in EJ movements to restore emotional as well as physical health to their damaged families and communities.[11]

Environmental Justice Is a Feminist Issue

According to the United Nations' *State of the World's Children* report released in January 2006, "Empowering women and eliminating gender discrimination is key to improving the survival and well-being of children worldwide . . . Gender equity is 'pivotal to human progress' because it produces a 'double dividend' of benefiting women and children, and is also crucial to the health and development of families, communities and nations" (quoted in Teotonio, 2006, p. A9). Women get involved in environmental justice because they are much more affected by poverty issues (Adamson, 2002), because they absorb the brunt of economic and social decline locally and worldwide (Hossay, 2006), and because of their perception of personal risk to themselves, their families, and their communities. While Prindeville (2004) describes women's involvement as a "politics of care" that reflects their multiple identities as mothers, workers, and community members,[12] MacGregor (2006) argues against such oft-repeated *maternalist* explanations, preferring to see women's involvement as an active and politicized form of ecological citizenship (a point to be taken up in more depth below).

Verchick (2004) argues that the EJ movement is a feminist movement but simply isn't recognized as such. He sites as evidence: (1) that EJ is focused on women's traditional care-taking roles in the family and community, and (2) that EJ relies on feminist legal theory and methods (p. 64). The EJ literature often describes women as the primary organizers in cleaning up and healing toxic communities (see, for example, Stein, 2002) as an extension of their role as family caregiver. At the same time, EJ organizations challenge white male privilege and unmask its bias by using contextual reasoning based on feminist law practices and consciousness-raising (Verchick, 2004). The feminist emphasis on contextual analysis and situational ethics reminds us that not all women and not all communities are the same or want the same thing. But, by the same token, as MacGregor (2006) persuasively argues, why don't we challenge the stereotypical role of women as caregivers to the world and instead argue for a more just and less gendered division of labour both at home and in the public realm? *That* would make the EJ movement a feminist movement.

Women told MacGregor that motherhood both enabled and constrained their political activism, and that they drew on "spiritual, cultural, political, and intellectual traditions, in addition to maternal ones" (p. 148). For some women, activist work is an escape from their everyday lives as mothers and homemakers. For others, activist work is an extension of their cultural beliefs – their duty to give back to their community – or of their profession. She notes that much of the literature also ignores the costs of women's activism, not the least of which is burnout and the way it compromises women's health and well-being.

Tarter (2002) argues that cancer is a feminist EJ issue because women are not equally protected against environmental hazards and because cancer rates are not evenly distributed. He cites Steingraber's argument that the parts of women most often affected by cancer – the breasts, ovaries, uterus – are often valued by society more than women's minds or even their lives, that women have less money and power and are more deferential to doctors because of socialization, that carcinogens can be passed on to infants through their mother's breast milk; and that women's cancers are under-researched relative to men's. As Plevin (2004) says, the womb and the home are no longer safe havens, but toxic sites of "corporate earnings" (p. 230). Tarter adds, as Steingraber certainly knows, that women's fatty tissues store carcinogens in ways the male body doesn't. For all these reasons, cancer and environmental justice are feminist issues. Tarter asks, why are we so silent about the link between environmental toxins and increasing rates of cancer, especially given that the World Health Organization acknowledges that 80 percent of cancers are most likely environmental? Are we overly attached to the consumer lifestyles that toxic industrial production has given us? No doubt. But perhaps women are also seen as expendable.

Emphasizing human rights, even women's rights, isn't necessarily the solution. Cornwall and Molyneux (2006) examined "the intersection of formal rights with the everyday realities of women in settings characterized by entrenched gender inequalities and poverty" (p. 1175). They found that some places have more than one legal system, and that cultural norms sometimes present quite formidable obstacles to realizing rights. They document how rights-based arguments are more often used to shore up economic claims or entitlements, like the rights of multinational corporations, than women's rights. This doesn't mean that we should abandon the struggle for women's human rights; it means that much of this work is being subverted by entrenched interests of one kind or another, and that we must be vigilant at unpacking assumptions and questioning which women and whose rights are being served.

Once again, MacGregor's 2006 book *Beyond Mothering Earth* suggests that, instead of seeing women's activism as an extension of their roles as mothers and caregivers, why not see it as an active expression of citizenship?

By doing so we destabilize the notion of the feminine or maternal as an essential identity of women and, in its place, allow for more democratic and more oppositional positions such that women's political identities and activism are not forever tied to the private sphere of care and maternal virtue. Moreover, instead of romanticizing women's capacity to care, MacGregor argues, it should be politicized. If we do not challenge these notions, we affirm sexist ideas about women's place in society. This is particularly dangerous for women in inegalitarian societies, where their burdens are already enormous and adding more responsibilities is both untenable and unjust. Essentialist discourse, she notes, also stands in the way of much-needed solidarity between feminist women and progressive men.

Although all of the women she spoke with see their activism as extending from "practices of care and concern for others," MacGregor argues that this sense of responsibility is part of female acculturation, rather than simply innate, and that women's burden of doing all the caring work is unfair (p. 197). For these reasons, she argues that feminist ecological citizenship "entails giving greater attention to the conditions necessary for active citizen participation: child care and eldercare, a more egalitarian division of labour at home, participatory processes that respect the time scarcity of people who juggle multiple roles, and public – as opposed to privatized and feminized – responsibility for human welfare and the quest for environmental sustainability" (p. 216).

Standing at a Crossroads

We are standing at a crossroads, with two apparent paths we might take. At the end of one path are prosperous gated communities, hellishly impoverished slums, and everyone in armed camps fighting over limited resources. At the end of the other is no utopia, but a spirited place in which we grapple with environmental problems by examining structural inequities and historical injustices such as racism, sexism, and predatory planning, and at the same time work to create healthy, equitable, and sustainable communities. Jane Jacobs (2004), in *Dark Age Ahead,* imagined the conflict over resources becoming ever more entrenched, particularly with those in the global North unwilling to give up their cars and their lavish consumption patterns. Continue on this path and we plunge into chaos and, ultimately, a global struggle – if not outright war – over limited resources. The only way to find our way onto the other path is by developing rejuvenated grassroots democracy movements both locally and globally.

So, how might we get onto that second path, toward environmental sustainability and justice, from here? I've culled suggestions from numerous sources (listed in parentheses following each strategy), added a gendered perspective (where none existed), and summarize these below to highlight the myriad ways we might build such a movement:

(1) *Use feminist participatory methods:* While feminist participatory methods are not a panacea (nor a substitute for structural change), they are essential for including marginalized grassroots women and their voices in the process of social change. Feminist participatory methods can build consciousness, confidence, and community, enabling women to take direct action on the environmental injustices that affect their lives, including the barriers to their full participation in political decision making and other spheres of public life. (Brulle and Pellow, 2005; Cable, Mix, and Hastings, 2005; MacGregor, 2006; Toronto Women's City Alliance, 2004; see also Maguire, 2001)

(2) *Foster resistance to hegemonic ideologies:* The corrosive power of racism and sexism, as well as other hegemonic ideologies, cannot be overestimated. We need to be vigilant to the ways in which words and ideas are used both to discredit women's claims and to induce passive and mindless consumerism. We need to actively resist and speak out whenever these are used to marginalize and silence women and other oppressed groups. We must demand that media – whether mainstream, alternative, or community-based – accurately document women's experiences and struggles in environmental in/justice, and demonstrate a commitment to gender equity. (Hossay, 2006; Simpson, 2002; Stein, 2002; Toronto Women's City Alliance, 2004)

(3) *Build cross-cultural collaborations:* Women need to organize cross-culturally in order to build a movement strong enough to create meaningful and lasting change. Only by bringing in the voices and energies of women capable of working across multicultural lines and across borders will it be possible to build a robust collaborative movement for systemic institutional change. We will need to build alliances with men as well, but equity cannot be achieved by silencing women in the name of an inclusive movement. (Adamson, 2002; Anthony, 2005; Khosla, 2003; MacGregor, 2006; Toronto Women's City Alliance, 2004)

(4) *Re-invest in community infrastructure:* Women's organizations in Canada and the vital services they provide have been severely eroded by government funding cutbacks. Gender-responsive budgeting could reverse these cutbacks. As Khosla (2003) has eloquently argued, women need secular publicly accessible spaces, and seek "public re-investment in income-security, social equality and justice in Canadian public policy" (p. 78). She recommends the establishment of cross-cultural women's social planning groups, with stable funding, to address the combined effects of poverty, racialization, and gender disparities.

(5) *Include cultural production:* Bringing in cultural production – including everything from community arts practices to fiction – can be a compelling and creative form of political action capable of redefining the nature of the movement, helping to involve marginalized women pre-

viously excluded from it, and strengthening our collaborations within it. (Adamson, Evans, and Stein, 2002; Stein, 2002.)

(6) *Envision sustainable women-positive communities*: The possibility for just, equitable, and sustainable communities that value women (as well as men) seems remote, given the violence, growing disparities, and environmental degradation common today. However, without a vision for an alternative future, we are lost. Women must play an active role in imagining new forms of living together equitably and in harmony with the earth and with each other, both locally and globally. Another world is possible only if we can imagine it and work together to create the social, economic, and political relations that will make it so (Agyeman, 2005).

We know what is needed in principle, but we also need to know what is possible strategically. Analysing the successes and failures of EJ protests in the United States, Toffolon-Weiss and Roberts (2005) conclude that there are two critical elements to consider. The first is political timing and whether or not it is an opportune time that can be exploited politically. In other words, is it a political priority or a high-profile issue at the time? If it is, then the best time to get involved is when policies are just being developed, rather than when they have already been set. The second consideration is whether or not the issue can be made attractive to outsiders and is likely, therefore, to gain outside allies. Here they caution that collaboration is easiest if winning doesn't require the displacement of your opponents. And that, I suppose, is the rub.

Global environmental issues, such as climate change, have the highest profile worldwide that they've ever had. If there were ever a time to create meaningful change in environmental policy, it would be now. Climate change and other global environmental problems are not affecting just impoverished women of colour, though they clearly do so disproportionately. The opportunity to make common cause with people across differences of race/ethnicity, class, gender, and nation is now. But what about those who have lived their whole lives in privileged spaces where their sense of entitlement has grown as large as their money can buy? They *will* have to be displaced; they *will* have to stop their excessive consumption and waste; and they *will* have to share their space, resources, and power.

Conclusion

Environmental justice and sustainability is a struggle. It must be a struggle that is highly participatory and includes marginalized women as key participants in the process. To create meaningful political, economic, and social change will require well-organized community-based networks with situated knowledge of local conditions and strong coalitions with similar organizations around the world. While the battle lines may be clearly demarcated in the United States, both socially and spatially, with long-standing communities of colour bound together by their tradition of struggle for civil rights and environmental justice, few Canadian communities have much shared history at all. Canadian cities, which have seen substantial changes in the ethnicity of immigrants over the past forty years, are also now witnessing increasing ethnic segregation. Depending on their varied experiences of settlement, many of these communities are just beginning to wake up to issues of EJ associated with the risks and hazards, as well as the lack of facilities and services, of their urban locations. Low-income and racialized women, in particular, seek to overcome their profound isolation and participate fully in public life.

Notes

1 Peggy Shepard of West Harlem Environmental Action, Inc., and chair of Summit II held in Washington, DC, October 2002, is quoted under the heading "Environmental Justice Summit Draws Over 1,200 Delegates" on the website http://www.ejrc.cau.edu/SummitII delegatenews.html. The summit included a Crowning Women Awards Dinner recognizing "sheroes."

2 EJ analyses also frequently neglect other marginalized groups, most noticeably queers, people with disabilities, and people living with AIDS. In the United States, Rachel Stein's 2004 edited volume, *New Perspectives on Environmental Justice: Gender, Sexuality, and Activism,* is a refreshing exception.

3 The concept of environmental space is credited to J.B. Opschoor, *Duurzaamheld en Verandering: over de Ecologische Ipastaarheid van Economische Ontwikkelingen [Sustainability and Change: On the Ecological Viability of Economic Development]* (Amsterdam: VU-Boekhandel, 1987).

4 See M. Wackernagel and W. Rees, *Our Ecological Footprint: Reducing Human Impact on the Earth* (Philadelphia: New Society, 1996).

5 See also B. Rahder and R. Milgrom, "The Uncertain City: Making Space(s) for Difference," *Canadian Planning and Policy – Aménagement et politique au Canada, 13*(1), 27-45.

6 Predatory planning is a concept developed by Kiara Nagel and defined as "an aggressive and deliberate practice of using land use zoning, public policy, law, and city planning to knowingly remove assets from the public or the poor to benefit a few or the very wealthy." Nagel, K.L., with J.E. Nagel, "Losing Our Commons: Predatory Planning in New Orleans; The Importance of History and Culture in Understanding Place," *Multicultural Review, 16*(1), 28-33; quote appears on p. 28.

7 For excellent archival footage, including interviews and newsclips, about the struggles of Africville residents, see the CBC digital archives website at http://archives.cbc.ca/IDC-1-69-96-471/life_society/africville/clip1.

8 See Statistics Canada. *2001 Census.* http://www12.statcan.ca/english/census01/home/index.cfm.

9 Ibid.

10 See V. Plumwood, *Feminism and the Mystery of Nature* (New York: Routledge, 1993), and E.K. Sedgwick, *Epistemology of the Closet* (Berkeley: University of California Press, 1990).

11 Stein (2002) shows how women's fiction can play a powerful role in drawing the links between environmental and ideological toxicity, and illustrate women's roles in healing the environment and their communities.

12 Although Prindeville uses the term *politics of care* to describe Hispanic mothers' motivation for getting involved, she notes that their activism is shaped by their multiple identities as mothers, workers, and community members.

References

Adamson, J. 2002. Throwing Rocks at the Sun: An Interview with Teresa Leal. In J. Adamson, M.M. Evans, and R. Stein (Eds.), *The Environmental Justice Reader: Politics, Poetics, and Pedagogy* (pp. 44-57). Tucson: University of Arizona Press.

Adamson, J., M.M. Evans, and R. Stein (Eds.). 2002. Introduction: Environmental Justice Politics, Poetics, and Pedagogy. In J. Adamson, M.M. Evans, and R. Stein (Eds.), *The Environmental Justice Reader: Politics, Poetics, and Pedagogy* (pp. 3-14). Tucson: University of Arizona Press.

Agyeman, J. 2005. *Sustainable Communities and the Challenge of Environmental Justice.* New York: New York University Press.

Agyeman, J., R.D. Bullard, and B. Evans (Eds.). 2003. *Just Sustainability: Development in an Unequal World.* Cambridge, MA: MIT Press.

Anthony, C. 2005. The Environmental Justice Movement: An Activist's Perspective. In D.N Pellow and R.J. Brulle (Eds.), *Power, Justice, and the Environment: A Critical Appraisal of the Environmental Justice Movement* (pp. 91-98). Cambridge, MA: MIT Press.

Brulle, R.J., and D. Pellow. 2005. The Future of Environmental Justice. In D.N. Pellow and R.J. Brulle (Eds.), *Power, Justice, and the Environment: A Critical Appraisal of the Environmental Justice Movement* (pp. 293-300). Cambridge, MA: MIT Press.

Buhrs, T. 2004. Sharing Environmental Space: The Role of Law, Economics and Politics. *Journal of Environmental Planning and Management, 47*(3), 429-47.

Cable, S., T. Mix, and D. Hastings. 2005. Mission Impossible? Environmental Justice Activists' Collaborations with Professional Environmentalists and with Academics. In D.N. Pellow and R.J. Brulle (Eds.), *Power, Justice, and the Environment: A Critical Appraisal of the Environmental Justice Movement* (pp. 55-75). Cambridge, MA: MIT Press.

Cornwall, A., and M. Molyneux. 2006. The Politics of Rights – Dilemmas for Feminist Praxis: An Introduction. *Third World Quarterly, 27*(7), 1175-91.

Dobson, A. 2003. Social Justice and Environmental Sustainability: Ne'er the Twain Shall Meet? In J. Agyeman, R.D. Bullard, and B. Evans (Eds.). *Just Sustainability: Development in an Unequal World* (pp. 83-95). Cambridge, MA: MIT Press.

Fong, E., and K. Shibuya. 2000. The Spatial Segregation of the Poor in Canadian Cities. *Demography, 37*(4), 449-59.

Gaard, G. 2004. Toward a Queer EcoFeminism. In R. Stein (Ed.), *New Perspectives on Environmental Justice: Gender, Sexuality, and Activism* (pp. 21-44). New Brunswick, NJ: Rutgers University Press.

Growth of Food Banks Signals Rise in Poverty. 2006, 10 December. *Toronto Star,* p. A16.

Gunder, M. 2006. Sustainability: Planning's Saving Grace or Road to Perdition? *Journal of Planning Education and Research, 26*(2), 208-221.

Hossay, P. 2006. *Unsustainable: A Primer for Global Environmental and Social Justice.* New York: Zed Books.

Inside the US Income Gap. 2007, 9 May. *Toronto Star,* p. B6.

Jacobs, D. 1992. Walpole Island: Sustainable Development. In D. Engelstad and J. Bird (Eds.), *Nation to Nation: Aboriginal Sovereignty and the Future of Canada* (pp. 179-85). Toronto: Irwin Publishing.

Jacobs, J. 2004. *Dark Age Ahead.* New York: Random House.

Kaalund, V. 2004. Witness to Truth: Black Women Heeding the Call for Environmental Justice. In R. Stein (Ed.), *New Perspectives on Environmental Justice: Gender, Sexuality, and Activism* (pp. 78-92). New Brunswick, NJ: Rutgers University Press.

Khosla, P. 2003. *If Low Income Women of Colour Counted in Toronto.* Toronto: Community Social Planning Council of Toronto.

Lucas, A.E. 2004. No Remedy for the Inuit. In R. Stein (Ed.), *New Perspectives on Environmental Justice: Gender, Sexuality, and Activism* (pp. 191-206). New Brunswick, NJ: Rutgers University Press.

MacGregor, S. 2006. *Beyond Mothering Earth: Ecological Citizenship and the Politics of Care.* Vancouver: UBC Press.

Maguire, P. 2001. Uneven Ground: Feminisms and Action Research. In P. Reason and H. Bradbury (Eds.), *Handbook of Action Research: Participative Inquiry and Practice* (pp. 59-69). London: Sage Publications.

Marcuse, P. 1998. Sustainability Is Not Enough. *Environment and Urbanization, 10*(2), 103-11.

McLaren, D. 2003. Environmental Space, Equity and the Ecological Debt. In J. Agyeman, R.D. Bullard, and B. Evans (Eds.), *Just Sustainability: Development in an Unequal World* (pp. 19-37). Cambridge, MA: MIT Press.

Ornstein, M. 2000. *Ethno-Racial Inequality in the City of Toronto: An Analysis of the 1996 Census.* Toronto: City of Toronto.

Pellow, D., and R.J. Brulle. 2005. Power, Justice, and the Environment: Towards Critical Environmental Justice Studies. In D.N. Pellow and R.J. Brulle (Eds.), *Power, Justice, and the Environment: A Critical Appraisal of the Environmental Justice Movement* (pp. 1-19). Cambridge, MA: MIT Press.

Plevin, A. 2004. Home Everywhere and the Injured Body of the World: The Subversive Humor of *Blue Vinyl.* In R. Stein (Ed.), *New Perspectives on Environmental Justice: Gender, Sexuality, and Activism* (pp. 225-39). New Brunswick, NJ: Rutgers University Press.

Prindeville, D.-M. 2004. The Role of Gender, Race/Ethnicity, and Class in Activists' Perceptions of Environmental Justice. In R. Stein (Ed.), *New Perspectives on Environmental Justice: Gender, Sexuality, and Activism* (pp. 93-108). New Brunswick, NJ: Rutgers University Press.

Richardson, M., J. Sherman, and M. Gismondi. 1993. *Winning Back the Words: Confronting Experts in an Environmental Public Hearing.* Toronto: Garamond Press.

Simpson, A. 2002. Who Hears Their Cry? African American Women and the Fight for Environmental Justice in Memphis, Tennessee. In J. Adamson, M.M. Evans, and R. Stein (Eds.), *The Environmental Justice Reader: Politics, Poetics, and Pedagogy* (pp. 82-104). Tucson: University of Arizona Press.

Stein, R. (Ed.). 2004. *New Perspectives on Environmental Justice: Gender, Sexuality, and Activism.* New Brunswick, NJ: Rutgers University Press.

–. 2002. Activism as Affirmation: Gender and Environmental Justice in Linda Hogan's *Solar Storms* and Barbara Neely's *Blanche Cleans Up.* In J. Adamson, M.M. Evans, and R. Stein (Eds.), *The Environmental Justice Reader: Politics, Poetics, and Pedagogy* (pp. 194-212). Tucson: University of Arizona Press.

Tarter, J. 2002. Some Live More Downstream than Others: Cancer, Gender, and Environmental Justice. In J. Adamson, M.M. Evans, and R. Stein (Eds.), *The Environmental Justice Reader: Politics, Poetics, and Pedagogy* (pp. 213-28). Tucson: University of Arizona Press.

Teelucksingh, C. 2002. Spatiality and Environmental Justice in Parkdale (Toronto). *Ethnologies, 24*(1). 119-41.

Teotonio, I. 2006, 11 December. Equity Key to Child Health: Report. *Toronto Star,* p. A9.

Toffolon-Weiss, M., and T. Roberts. 2005. Who Wins? Who Loses? Understanding Outcomes of Environmental Injustice Struggles. In D.N. Pellow and R.J. Brulle (Eds.), *Power, Justice, and the Environment: A Critical Appraisal of the Environmental Justice Movement* (pp. 77-90). Cambridge, MA: MIT Press.

Toronto Women's City Alliance. 2004. A New Deal for Women: What Women in Toronto Need. http://www.twca.ca/publications.php.

Unger, N. 2004. Women, Sexuality, and Environmental Justice in America. In R. Stein (Ed.), *New Perspectives on Environmental Justice: Gender, Sexuality, and Activism* (pp. 45-60). New Brunswick, NJ: Rutgers University Press.

United Way of Greater Toronto and the Canadian Council on Social Development. 2004. *Poverty by Postal Code: The Geography of Neighbourhood Poverty 1981-2001.* Toronto: United Way of Greater Toronto and the Canadian Council on Social Development.

Verchick, R. 2004. Feminist Theory and Environmental Justice. In R. Stein (Ed.), *New Perspectives on Environmental Justice: Gender, Sexuality, and Activism* (pp. 63-77). New Brunswick, NJ: Rutgers University Press.

Walks, R.A., and L.S. Bourne. 2006. Ghettos in Canada's Cities? Racial Segregation, Ethnic Enclaves and Poverty Concentration in Canadian Urban Areas. *Canadian Geographer, 50*(3), 273-97.

5

The Political Economy of Environmental Inequality: The Social Distribution of Risk as an Environmental Injustice

S. Harris Ali

In recent years, environmental justice groups have arisen to deal with those issues related to the inequitable distribution of the "externalities" of industrial production. Many of the matters dealt with by these groups, as well as the analyses of the environmental justice movement itself, have tended to have a decidedly *local* focus – dealing with specific community-contamination events, the harmful health effects of local pollution on members of the community, and the local politics involved therein. This is understandable given that environmental justice issues are very much place-based and context dependent (Parizeau, 2006). At the same time, however, what is often neglected in analyses, as David Pellow (2000) points out, is the question of the structural environmental inequality that gives rise to the unjust distribution of environmental risks in the first place. Addressing this important foundational issue requires a broader focus on the intersection of environmental quality and social inequality, which in turn demands a more process-oriented perspective sensitive to larger sociohistorical processes that inform and influence the accumulation of externalities in a particular place. Without this analytical reorientation toward a more critical understanding of structural inequality, the literature that supports the environmental justice movement will continue to have a limited theoretical perspective that leads to description rather than explanation (Parizeau, 2006). In this chapter, I respond to this call for a reorientation by considering how the broader forces involved in the Canadian political economy expose those in particular regions to harm. This analysis of the social distribution or risk will be illustrated by drawing upon the experience of coal-mining disasters and toxic contamination from steelmaking in Nova Scotia.

The Political Economy of Canada

Much of the work on the Canadian political economy can be traced in some way or another to the pioneering work of the economic historian

Harold Innis (1954) on staples theory. Innis argues that Canadian economic development was largely shaped by the explicit adoption of a trade policy that emphasized the flow of natural resource staples such as fur, fish, wheat, coal, metals, and timber from rural communities (i.e., hinterlands) to urban centres (i.e., metropoles) where they would be processed and/or used in the manufacturing of commodities. In essence, the resultant staples economy was based on the trade of staples from the colonized for manufactured goods from the colonizer. Thus, the initial role of Canada as a white settler colony was to supply cheap food and primary resources to Britain, and as such, the major focus of investments was on staples extraction and not industrial manufacturing (Williams, 1994).

For some of the earlier scholars, including W.A. Mackintosh, a staples-based economy was optimistically construed as merely a transitional stage that would soon give way in the future to a more broadly based industrial society (Schedvin, 1990). Others, including Innis himself, were less optimistic, believing that natural resource communities would increasingly become economically dependent on the metropoles located in areas quite distant from the hinterland in which the staples originated. This was believed to cause problems for the hinterland communities because of the periodic instabilities that arose from an unregulated market for staples. Furthermore, this precariousness would not allow the economy to mature or establish itself in any stable or permanent manner. Natural resource-dependent towns would therefore go from one crisis to the next with the closing of mills and mines, the loss of jobs, and the outmigration of residents.

Historically, the reliance on staples has influenced the direction and nature of Canadian economic development in various ways. Most notably, this has included an overemphasis on the building of infrastructure to support staples-based industries. And as Bunker (2005, p. 39) notes, infrastructure development involving roads, rails, and ports generally involve large amounts of very inflexibly invested capital. Consequently, natural resource economies tend to develop a "spatial fix" (Harvey, 1982) in which high levels of capital inflexibility are sunk in technologies of extraction and export. For example, the construction of the railroad across Canada was a priority for the country's first politicians because it was thought that the railway would unify the country by facilitating the transport of natural resources across vast distances to metropole markets. Innis notes, however, that the establishment of transportation infrastructures was very costly and led to the incurring of heavy debts to foreign lenders, which further intensified the economic dependence of Canada on foreign interests, particularly those in the United States and Britain. Furthermore, such a strategy has meant that resource extraction was overemphasized at the expense of industrial expansion in manufacturing (Williams, 1994). For these reasons, it has been argued that Canada had failed to develop a substantial trade in finished manufactured goods while retaining an export structure based on primary resources – a pattern actually found in many underdeveloped nations of the world. Moreover, because primary resource industries tend to be capital-intensive rather than labour-intensive (as in manufacturing), Canada has historically sacrificed its potential for job creation (Williams, 1994). Many of these developments were supported by the National Policy of the country's first prime minister, who sought to promote trade between the provinces rather than trade between Canada and other countries. This protectionist strategy was largely pursued through the imposition of heavy tariffs on goods that were to be imported into Canada, thus serving as a disincentive for other countries, the United States among them, to trade with Canada.

The legacy of the staples economy is still found in the nature of the uneven development found in Canada. In more recent years, this is seen in terms of the effects of adopting an industrial strategy known as import substitution industrialization. With this strategy, the reliance on exports is severed by promoting the trade of manufactured commodities and primary resources within the country itself – that is, between different regions of the country. Because of the lack of domestically owned industry in Canada, those commodities that are made here are manufactured on the basis of technologies, equipment, and funds supplied by sources outside the country. In the case of Canada, much of the direct foreign investment was made by American capital (Clement, 1983). Moreover, to avoid the heavy tariffs on American goods, American capitalists would establish branch plants in Canada so that American technology would be used to produce goods on Canadian soils. However, the profits were not reinvested in the local Canadian economy but funnelled back into the United States, thus reinforcing the dependent relationship the Canadian hinterland had with the US metropole in the form of a branch-plant economy (Williams, 1994).

Regional Disparity and Regional Dependence

Much of the above discussion of the Canadian political economy focuses on the dependency relationship between Canada and first Britain, then later the United States. The legacy of the staples economy, the National Policy, and import substitution have also led to unique forms of dependency relationships *within* Canada – that is, between different regions or provinces. The nature of these relationships is tied to the role of natural resources in Canada. Although most resources are provincially owned, their control for the large part has been given to private firms. Consequently, resource development may be hindered or prevented by the corporate policies of private firms, instead of being under the sole direction of the government. It is precisely at this critical juncture that we begin to see how the social, occupational, and environmental risks associated with primary resource extraction and processing arise, and how such origins influence the subsequent environmental justice politics that surround these problems.

In considering disparities between different areas within the country, it is useful to consider the region as the unit of analysis. In this light, central Canada (primarily Toronto and Montreal) is a metropole region, while the western and Atlantic provinces are the hinterland regions that supply the central region. Before Confederation in 1867, the Maritime region was economically prosperous, as those provinces were engaged in a vibrant trade in fish, timber, and ships with England, the United States, and the West Indies (Matthews, 1983). At the same time, central Canada was at a disadvantage because it was cut off from the sea during the winter months when the St. Lawrence River was frozen. With the advent of railway construction, however, the commercial elite of Montreal and Toronto consolidated their positions in banking policy, railway policy, and tariff polices, all of which ensured increased wealth for those in economic and political power in central Canada.[1] This competitive advantage for central Canada was supported by state policy following Confederation that made the establishment of local banking difficult in the Maritimes (Clement, 1983, p. 64). Banks in central Canada were able to ensure a shortage of loan capital and higher interest rates, thereby undercutting the ability of potential Maritime manufacturers to compete. Thus, the possibility of establishing a thriving manufacturing base in the Maritime region was thwarted and the region reverted to staples production. Furthermore, the lack of an established manufacturing and industrial base led to high unemployment rates in the Maritime region. Such broad political-economic machinations related to the emergence of regional disparities and ultimately established the circumstances for the creation of hazardous environmental health conditions in Nova Scotia, particularly in the realms of coal mining and steel production.

Coal-Mining Disasters in Nova Scotia

During the course of mining coal in Nova Scotia, there have unfortunately been many disasters, with tragic losses of life. I will focus on just two incidents here to illustrate how broader political-economic forces play a significant role in making certain communities and occupational groups within particular regions vulnerable to hazards. On 23 October 1958, a catastrophic "bump" – the sudden shifting of underground strata that leads to the collapse of tunnels – occurred in a mine owned by the Dominion Steel and Coal Corporation (DOSCO) in Springhill, Nova Scotia. Of the 174 miners who were trapped, 75 died and 99 were rescued. Of those who survived, miraculously, 12 of them crawled out of the mine six days after the bump, while another 7 did so three days after that (Greene, 2003).[2]

Although the cause of the disaster was popularly attributed to be the bump itself, McKay (1987, p. 165) notes that such a simplistic explanation is misleading, as the "real" causes of the disaster are implicated in the history of the region's coal industry and the way decisions are made in the workplace. Thus, a broader examination is required.

With the decline in the amount of coal required by the Canadian National Railway in the 1950s (as its trains switched to diesel), the coal industry faced tough economic times and so turned toward both provincial and federal governments for subventions and conventions (McKay, 1987, p. 166). It was also during this period that the community lost its power to effectively bargain with DOSCO and the government because of threats of company relocation. Miners for a long time warned of the unstable conditions of the Springhill mine, but their warnings went unheeded by management, despite the frequent recurrences of smaller scale bumps. Through various public meetings with DOSCO, the members of the mining community expressed their concerns of the company's plan to keep digging the town's only operating mine further and deeper despite the increased risk of a cave-in such digging entailed. They also insisted that the company reopen abandoned mines in the area that were known to still have significant levels of coal and that were safer. The company rejected this suggestion because it was much more costly for it to reopen those mines than to continue digging at Springhill. The dominance of coal mining in the region led to a general de-emphasis on educational attainment, as this was not required for the occupation. Members of the region's labour force were therefore less able to switch jobs – if any did exist in their resource-based communities in the first place (Tupper, 1978). The miners, therefore, were forced to accept the risks they knew to exist in the Springhill mine in order to gain some livelihood in the harsh economic realities they faced.

The negative economic circumstances faced by those in the Atlantic region in the 1950s has continued into even more recent times, with similar tragic results. On 9 May, 1992, a methane explosion in the Westray mine in Plymouth, Nova Scotia, killed twenty-six miners. As with the Springhill disaster, warnings of the imminent dangers of the coal-mining practices were ignored by management. This reflected a long history of a lack of concern for abiding by industry regulations. In fact, the health and safety laws in Nova Scotia lagged behind that of other provinces – a consequence of the pressures stemming from regional dependence (as we shall see). Furthermore, during the 1990s, the fines for violations of the provincial act governing coal mining in Nova Scotia were often nominal (e.g., $250), with sanctions rarely imposed in any case (Tucker, 1995). The regulatory system for occupational safety in the region was also lax. For example, from 1985 to 1990, although fourteen mining companies were charged for violation of the provinces' occupational health and safety act, none of them was prosecuted. The lack of attention to the health and safety of workers is often understood to occur because of a natural resource company's drive to maximize profit, but what accounts for the lack of government attention to health and safety? For example, Tucker (1995) asks, why did the government not require occupational health and safety assessments to be performed before granting approval for or financial assistance to mines? For Tucker (1995), the answer lies in the emphasis on regional development at all costs in Nova Scotia. Such a conclusion is supported in no uncertain terms by an inquiry review of an explosion at a Cape Breton coal mine in 1979 that led to the death of twelve miners: "The social and industrial expectations and acceptance of unnecessary risks over many years against the possible loss of employment had fostered attitudes and environmental conditions that made this explosion and previous fires almost inevitable . . . the production of coal was given a priority over almost all other considerations" (Canada, 1979, cited in Tucker, 1995). Again, unsafe and unlawful practices were imposed on workers who did not raise objections out of fear of losing their jobs or retaliation by management.

McMullan (1997) notes that the certain environmental harms – those discussed in this chapter – may be considered as a specific type of organizational misconduct that Kramer and Mischalowksi (1990, p. 3) refer to as a state-corporate crime. Such crimes result from the cooperation between governments and private companies. Regional development policies in Canada have tended to encourage state-corporation relationships in ways different from that of United States. Whereas the United States emphasizes a strong adherence to a noninterventionist approach to the economy, Canada does not. For example, in 1967, the federal government established the Crown corporation Cape Breton Development Corporation (CBDC) to solve the province's coal "problem" stemming from (1) the region's dependence on DOSCO for employment and (2) the coal industry's chronic economic instability (Tupper, 1978). The Crown corporation was actually formed in response to DOSCOs' announcement that it would close its unprofitable coal mines in the region. With the introduction of the Crown corporation, the coal industry would be run as a state enterprise, with the commendable intentions of ending the thoughtless exploitation of the regional economy by private enterprise – DOSCO was the region's largest employer (McKay, 1987, p. 165). It was hoped that feelings of exploitation would be replaced with the sentiment of working for one's self and one's community (Tupper, 1978). It was also thought that state ownership would enable the gradual phase-out of unprofitable mines and reduce unemployment slowly, while allowing time for the regional economy to diversify (Tupper, 1978). However, state ownership of coal mining (and later steel production) had the effect of intensifying the environmental hazards members of the community faced. For example, because the coal mined in the region was of an impure type, containing various contaminants (such as sulphur), the resultant steel was of relatively poor quality. Consequently, DOSCO had begun to use coal imported from Pennsylvania – this was both cheaper and of better quality. When taken over by the Crown corporation, the steel company again started to use the inferior coal mined in Cape Breton. The problem was that the use of poor-quality coal resulted in greater environmental contamination. Several studies conducted by the federal government in the 1970s revealed that the increased air pollution from the impure coal was having a statistically significant detrimental impact on public health in the region – although these reports were suppressed from the public (Barlow and May, 2000).

Toxic Contamination from Steelmaking: The Case of the Sydney Tar Ponds

The harbour city of Sydney, Nova Scotia (population: 26,872), is located on the northeast corner of Cape Breton Island. The abundant coal here was used to produce coke that could in turn be used in steelmaking, and in 1899, the Dominion Iron and Steel Company (DISCO) was established on Sydney Harbour by an American interest, a Bostonian named Henry Whitney. Notably, DISCO received numerous government concessions as inducements to establish itself in Sydney, including a land grant of two hundred hectares (some of which was appropriated without compensation from the indigenous Mi'kmaq peoples) (Barlow and May, 2000, p. 8), a thirty-year tax holiday, and a special rate for water and coal (Campbell, 2002; Crawley, 1990). Over the proceeding years, there were various mergers between the steel plant and the coal operations, with accompanying name changes and various owners from Montreal and Toronto holding the controlling interest (Abbass, 2006). Many of these new owners were also involved in Canadian Pacific Railway companies and, in fact, much of the steel produced in Sydney was used in the manufacture of rails. In 1957, the steel operation, which eventually was renamed the Dominion Steel and Coal Company (DOSCO, the same company that owned the Springhill mine), was sold to the British multinational Hawker Siddeley. In 1967, the British interests wanted to close down the operations because of the poor-quality steel that was produced (and difficult to sell) and the world slump in the international steel market. This led to a widespread protest in Sydney, with residents marching in the streets on 13 October 1967 to demand that the plant continue operations (Abbass, 2006).

In response to the public outcry, in December 1967 the Province of Nova Scotia formed a Crown corporation to take over DOSCO, under the new name of Sydney Steel Corporation (SYSCO).[3] For the next thirty-three years, various modifications were made in the steel-production process to make the steel facility economically competitive, but for the most part these attempts failed. As Barlow and May (2000) note, very little attention or money was spent on improving the company's occupational and environmental health record during this time of state ownership. Under state control, the levels of pollution and contaminants increased substantially, while efforts to curb these were not even considered. For example, the 150 tons of toxic blast furnace dust that coated areas of Sydney each month could have easily been curtailed with the installation of a $6 million cintering plant – a small sum for a large company such as SYSCO – but this was never seriously considered either by the company itself or at the government's insistence, despite the significant public health benefits it would have for the community (Abbass, 2006, p. 17; Barlow and May, 2000, p. 17).

Over the many years of operation, slag waste from the blast furnace, as well as other toxic by-products from the steelmaking and coke operations, collected in the estuary leading to the harbour. Much of the estuary became completely landfilled over time, with mountains of accumulating slag forming a barrier that ran through the estuary. Pools of toxic material collected in the estuary and came to be known as the "tar ponds"; daily current flows brought the toxic materials from the tar ponds into the harbour itself. Despite numerous government-sponsored health studies (kept hidden from the public in SYSCO offices) revealing higher than expected cancer rates in the area, as well as other significant threats to public health, the steel plant was allowed to continue to pollute until 1980. It was at this point that the Department of Fisheries discovered that lobsters in the harbour contained such high levels of toxic contaminants, particularly PCBs, that they were not fit for human consumption (Barlow and May, 2000, p. 75). Consequently, fishing was prohibited in some parts of the harbour, and public concern over the health impacts associated with the tar ponds increased.

Government attention now began to focus on cleaning up the tar ponds. Various toxic assessment and remediation initiatives were performed by technical consultants. It was soon understood that the site was one of the most contaminated and toxic areas in the country. Government officials decided that the incineration of toxic material would be pursued and so, in 1991, the provincial Crown corporation of Sydney Tar Ponds Clean-Up Inc. was formed to own and operate the incinerator. However, certain technological design flaws became evident as operations began. After several costly ($55 million) and repeated attempts to the fix these flaws, and the finding of much higher-than-anticipated levels of PCBs in the toxic waste, the incineration plan was abandoned (incineration cannot degrade PCBs). The Nova Scotia government then decided that the maximum amount it was prepared to spend on the cleanup was $20 million – not because that was the estimate of the costs of cleanup but because that was how much money the provincial government thought it could afford to spend (Barlow and May, 2000, p. 88). Without inviting tenders, the government approached the Nova Scotia engineering firm of Jacques Whitford to see what could be done for that amount of money. The firm suggested a process of encapsulation whereby the mountains of slag would be used to fill in the tar ponds, then capped with soil and grass to produce a park. When this announcement was made, the Sydney community reacted with outrage, especially since it had not been consulted. Ultimately, a federal ruling overturned this decision to cap the contaminated area. The provincial government then moved to adopt a more open consultation process and announced the formation of a new community-government committee with a budget of $1.67 million. It would involve three levels of government and the Sydney community working together to form a new cleanup plan known as the Joint Action Group (JAG) process (Campbell, 2002). JAG membership consisted of government representatives and a roundtable open to the public. Fifty-five community members joined JAG and numerous working subcommittees were quickly formed to deal with various issues such as those involving health studies, site security, remediation options, planning, governance, human resources, finances, and ethics. But the manner in which the JAG process unfolded did not allow for the opportunity to deal with larger issues related to environmental inequality and environmental justice.

Implications of Conceptualizing the Social Distribution of Risk as an Environmental Injustice

It has been argued that in order to achieve more penetrating and critical insights into local environmental justice issues, the breadth of focus must be broadened to include issues of structural and environmental *inequality*. In this light, I have attempted to demonstrate how such political-economic dynamics operating at the national and regional level have repercussions for the state of the environment and issues of environmental justice vis-à-vis the distribution of risks. In this sense, localized particular injustices can be viewed as the consequences of specific constellations of power relations (Parizeau, 2006). Thus, for example, we have seen how decisions made by those in central Canada (such as Toronto and Montreal) and foreign owners (United States and Britain) have influenced the nature of the staples economy, and ultimately the nature of future Canadian development. This has led to situations of regional dependence and regional disparity that reflect the social distribution of risks, as, for example, with the risks of coal mining and steelmaking accumulating in certain regions, such as Cape Breton Island and Nova Scotia. Furthermore, a focus on regional disparity reveals how the injustices experienced by those in this region were limited not only to the unjust exposure to environmental risks but also to the social and economic exclusion that distances those in Atlantic Canada from the more prosperous populations in other regions (Social and Economic Inclusion Initiative, 2003, cited by Haalboom, Elliott, Eyles, and Muggah, 2006, p. 240). As such, environmental inequality reflects social and economic inequality. The practical question still remains: What does an emphasis on the broader structural aspects of environmental inequality mean for the analysis and practice of environmental justice?

Perhaps most significantly, a focus on the broader context of environmental inequality will enable both the analyst and the activist to understand *community vulnerability* as an environmental justice issue and to design appropriate strategies thereof. Community vulnerability does not mean only the susceptibility of particular social groups to technological and environmental disasters such as chemical contamination that have traditionally been the types of struggles with which environmental justice groups have been involved. It also includes the social and economic vulnerabilities tied to a community's dependence on natural resource extraction and primary industries – after all, natural resource issues are obviously environmental issues as well. As such, as Hessing (2002, p. 36) notes, the ecological consequences of economic restructuring requires special consideration within the Canadian context because many communities remain directly dependent on resource availability. In addition, MacLeod, McFarlane, and Davis (1997) note that natural resource towns may quickly become distressed communities characterized by high unemployment, a loss of vital services (schools, hospitals, housing, finance), the deterioration of transportation and communications infrastructure, the loss of population, and notably, a declining influence in central political institutions. Such conditions are triggered when a single-industry town loses its principal source of income, such as when a mine or steel mill closes. Two things are achieved by emphasizing community vulnerability as a product of environmental inequality. First, it better enables connections to be made between environmental health issues and social justice issues. Second, it will help counter the common tendency of risk management officials to individualize the risks – that is, to blame the victim. For example, in response to initial health reports on the higher cancer rates among Sydney residents, government officials claimed that such rates were not because of environmental contamination from the coke ovens and steelmaking but because of individual lifestyle factors, such as excessive smoking, drinking, and poor diet (Barlow and May, 2000, p. 145). Another individualizing tendency was seen in the strategy of public officials to dismiss anecdotal evidence of cancer deaths in the community as a general conspiracy to stop steel production (Rainham, 2002). Furthermore, it was evident that local politicians were more concerned that the environmental health issues raised by citizens would deteriorate the value of land and housing prices, and, for that reason, citizens were frequently labelled as individual extremists who lacked any sort of credibility (Rainham, 2002). By emphasizing the community exposure to environmental harms, such individualizing claims may at least start to be countered through critical social epidemiology and popular epidemiological methods based on local knowledge and the critique of conventional reductionist techniques of the public health establishment (see, for example, Lambert, Guyn, and Lane, 2006, with reference to the Sydney Tar Ponds; and, more generally, Brown, 1992, 1997; Brown and Mikkelsen, 1990; Gibbs, 1995; Ali, 2002).

Often in environmental justice struggles the larger structural matters related to environmental inequality remain obscure and therefore the relevant state and economic actors remain free from criticism. Why then do such matters remain obscure? How does this political neutralization process occur? A study that included interviews with those involved in the Joint Action Group (JAG) process related to the Sydney Tar Ponds (Burke, 2007) found that much time and attention was focused on such things as personal conflicts between members and learning about technical issues related to remediation. Such attention diverts resources away from the larger structural questions.[4] For example, although studies had found that a certain community directly downwind from SYSCO, the Whitney Pier neighbourhood, was most affected by high levels of contaminants, during the JAG process the fact that Whitney Pier was largely a black and immigrant community was not directly raised (May, 2002). Nor was the other environmental racism issue of the forced historical resettlement of the Mi'kmaq to accommodate the steel and coke operations. Rather, the attention to risk management and risk communication that formed the basis of much of the JAG deliberations tended to narrow the focus and divert questions away from the unjust political-economic decisions that led to the placement of certain people in harm's way and the forced displacement of others. This confirms the findings of Hessing who found that "while many environmental groups seek access to the decision-making process, in practice they have little substantive influence on bilateral relations between state and private interests, which control amounts and conditions of resource exploitation" (2002, p. 39). In sum, the cases reviewed here reveal how potential critiques of the development policies that gave rise to the uneven social distribution of risk were circumvented by focusing exclusively on localized ad hoc technical matters. In this way, the underlying basis for environmental inequality remained intact and important structural causes for the inequality remained obscure and therefore not discussed.

Finally, some questions remain about how conceptualizations of environmental inequality in terms of dependency theory fare in the light of contemporary processes of globalized neo-liberalism. Under neo-liberal strategies, new means of accumulation and social regulation arise as authority from the public sphere (where it is subject to collective claims or debate) is partially transferred to the private domain – to the corporation, the community, or individuals. According to Young and Matthews (2007), neo-liberal reforms in resource-based economies have the objective of liberating major corporate actors from non-market obligations, particularly with respect to the environment, labour, and communities. This was indeed what happened in Nova Scotia, as the issues related to addressing problems of environmental remediation were no longer the responsibility of the private steel and coal corporation, nor was it in the hands of the state. The state devolved responsibility onto the community through the JAG process – hence the common complaint by JAG participants about the lack of government involvement in the process. As Haalboom et al. (2006) observe, the government was able to sidestep blame for the problems that arose in the Sydney tar ponds environmental remediation process because the JAG was considered an accountable agency, hence it was that agency that would receive the brunt of the public animosity rather than the government per se.

Conclusion

Many of the injustices addressed (or not) by local environmental justice groups have their origins in long-standing and entrenched structures of social and environmental inequalities. These inequalities do not arise in a vacuum but from broader political-economic forces that exert their influences at the local level and result in many types of environmental health threats. To deal more directly with the *source* of these threats, rather than only their effects, it is critical that environmental justice groups be aware of the historical context in which their specific issues originate and persist. In particular, by understanding the political-economic dynamics behind specific environmental justice issues, more effective and longer lasting strategies may be developed not only to prevent future environmental threats for those involved in a particular local campaign but to help out those in other localities within a region experiencing similar environmental injustices. In other words, it is important to take seriously the idea that environmental justice issues are particular instances of a broader pattern of environmental inequality, so that the frame is expanded in such a way as to encompass the environmental injustices of all those in a similar situation, rather than individualizing the problem to a locality. By addressing issues of environmental inequality in this way, we may slowly move toward the promise of environmental justice for all.

Notes

1 Nevertheless, central Canada itself still did not directly engage in industrialization – a void filled by the establishment of American branch plants to tap the major market there and because this region was closer to American operations in upper New York state (Matthews, 1983).

2 This story, which occurred in the early days of television, received a great deal of coverage, particularly on American television.

3 Recall similar actions had taken place a few months earlier in relation to coal mining with the federal government's establishment of the Cape Breton Development Corporation.

4 Members did themselves identify a lack of government involvement in the JAG process as a principle deficiency in the process (Burke, 2007), and it can only be surmised that had there been such involvement, at least the opportunity could have arisen to raise the larger structural issues in a critical fashion.

References

Abbass, F. 2006. *DISCO, DOSCO, SYSCO, 1672-2005.* Sydney, NS: Sunday Novel Publishing.

Ali, S.H. 2002. Dealing with Toxicity in the Risk Society: The Case of the Hamilton Ontario Plastics Recycling Fire. *Canadian Review of Sociology and Anthropology, 39*(1), 29-48.

Barlow, M., and E. May. 2000. *Frederick Street: Life and Death on Canada's Love Canal.* Toronto: HarperCollins.

Brown, P. 1997. Popular Epidemiology Revisited. *Current Sociology, 45*(3), 137-56.

–. 1992. Toxic waste contamination and popular epidemiology: Lay and professional ways of knowing. *Journal of Health and Social Behavior, 33,* 267-81.

Brown, P., and E.J. Mikkelsen. 1990. *No Safe Place: Toxic Waste, Leukemia, and Community Action.* Berkeley, CA: University of California Press.

Bunker, S.G. 2005. How Ecologically Uneven Development Put The Spin On The Treadmill of Production. *Organization and Environment, 18*(1), 38-54.

Burke, A. 2007. A Model for Recommendations in Community Engagement and Remediation Issues (MA thesis, York University, Toronto).

Campbell, R.A. 2002. A Narrative Analysis of Success and Failure in Environmental Remediation: The Case of Incineration at the Sydney Tar Ponds. *Organization and Environment, 15*(3), 259-77.

Canada. 1979. *Report of Commission of Inquiry, Explosion in No. 26 Colliery Glace Bay, Nova Scotia on February 24, 1979.* Ottawa.

Clement, W. 1983. *Class, Power and Property: Essays on Canadian Society.* Agincourt, ON: Methuen.

Crawley, R. 1990. Class Conflict and the Establishment of the Sydney Steel Industry 1899-1904. In K. Donovan (Ed.) *The Island: New Perspectives on Cape Breton History, 1713-1990* (pp. 145-64). Sydney: University College of Cape Breton Press.

Gibbs, L.M. 1995. *Dying from Dioxin: A Citizens' Guide to Rebuilding Our Health and Reclaiming Democracy.* Boston: South End Press.

Greene, M.F. 2003. *Last Man Out: The Story of the Springhill Mine Disaster.* Orlando: Harcourt.

Haalboom, B., S.J. Elliott, J. Eyles, and H. Muggah. 2006. The Risk Society at Work in the Sydney "Tar Ponds." *Canadian Geographer, 50*(2), 227-41.

Harvey, D. 1982. *The Limits of Capital.* Oxford: Blackwell.

Hessing, M. 2002. Economic Globalization and Canadian Environmental Restructuring. In S.C. Boyd, D.E. Chunn, and R. Menzies (Eds.), *Toxic Criminology: Environment, Law and the State in Canada* (pp. 25-43). Halifax: Fernwood.

Innis, H.A. 1954 [1940]. *The Cod Fisheries: The History of an International Economy.* New Haven, CT: Yale University Press. (Rev. ed.) Toronto: University of Toronto Press.

Kramer, R.C., and R. Mischalowksi. 1990. State-Corporate Crime. Paper presented at the annual meeting of the American Society of Criminology, Baltimore.

Lambert, T.W., L. Guyn, and S.E. Lane. 2006. Development of Local Knowledge of Environmental Contamination in Sydney, Nova Scotia: Environmental Health Practice from an Environmental Justice Perspective. *Science of the Total Environment, 368,* 471-84.

MacLeod, G., B. McFarlane, and C.H. Davis. 1997. The Knowledge Economy and the Social Economy: University Support for Community Enterprise Development as a Strategy for Economic Regeneration in Distressed Regions in Canada and Mexico. *International Journal of Social Economics, 24*(11), 1302-24.

Matthews, R. 1983. *The Creation of Regional Dependency.* Toronto: University of Toronto Press.

May, E. 2002. Frederick Street Revisited. In S.C. Boyd, D.E. Chunn, and R. Menzies (Eds.), *Toxic Criminology: Environment, Law and the State in Canada* (pp. 87-96). Halifax: Fernwood.

McKay, I. 1987. Springhill 1958. In G. Burrill and I. McKay (Eds.), *People, Resources, and Power: Critical Perspectives on Underdevelopment and Primary Industries in the Atlantic Region* (pp. 162-85). Fredericton: Acadiensis Press.

McMullan, J.L. 1997. State/Corporate Crime and Social Justice: Reflections on Politics, Power and Truth. Paper presented at the Conference on Social Justice, Social Inequality and Crime, Fredericton. http://www.stthomasu.ca/~ahrc/conferences/mcmullan.html.

Nicholls, W.J., and J.R. Beaumont. 2004. The Urbanisation of Justice Movements? *Space and Polity, 8*(2), 107-17.

Parizeau, K. 2006. Theorizing Environmental Justice: Environment as a Social Determinant of Health. In J.C. Cohen and L. Forman (Eds.), *Munk Centre for International Centre Briefings, Comparative Program on Health and Society Lupina Foundation Working Papers Series* (pp. 101-28). Toronto: University of Toronto.

Pellow, D.N. 2000. Environmental Inequality Formation: Toward a Theory of Environmental Justice. *American Behavioral Scientist, 43*(4), 581-601.

Rainham, D. 2002. Risk Communication and Public Response to Industrial Chemical Contamination in Sydney, Nova Scotia: A Case Study. *Journal of Environmental Health, 65*(5), 26-32.

Schedvin, C.B. 1990. Staples and regions of Pax Britannica. *Economic History Review, 43*(4), 533-59.

Social and Economic Inclusion Initiative. 2003. *Closing the Distance.* Toronto: Social Planning Network of Ontario. http://www.ccsd.ca/events/inclusion/papers/spno-initiative.pdf.

Tucker, E. 1995. The Westray Mine Disaster and Its Aftermath: The Politics of Causation. *Canadian Journal of Law and Society, 10*(1), 91-123.

Tupper, A. 1978. Public Enterprise as Social Welfare: The Case of the Cape Breton Development Corporation. *Canadian Public Policy, 4*(4), 530-46.

Young, N., and R. Matthews. 2007. Resource Economies and Neoliberal Experimentation: The Reform of Industry and Community in Rural British Columbia. *AREA: Journal of the Royal Geographical Society, 19*(2), 176-85.

Williams, G. 1994. *Not for Export: The International Competitiveness of Canadian Manufacturing.* (2nd ed.). Oxford: Oxford University Press.

6

These Are Lubicon Lands: A First Nation Forced to Step into the Regulatory Gap

Chief Bernard Ominayak, with Kevin Thomas

The Lubicon Lake people are an indigenous nation of approximately five hundred people living in northern Alberta. We have never surrendered our rights to our traditional territory in any legally or historically recognized way. When a treaty was negotiated with other indigenous peoples in the region in 1899, treaty negotiators did not travel inland to our territory and we were therefore left out of the treaty process. In the past twenty-seven years, our traditional territory has been invaded and ravaged by multi-billion-dollar resource exploitation activity, including logging and large-scale oil and gas extraction. These massive resource exploitation activities have decimated the traditional Lubicon hunting and trapping economy and way of life and threaten the very existence of the Lubicon Lake people as a distinct indigenous society.

Between 1979 and 1983, more than four hundred oil and gas wells were drilled within a 25-kilometre radius of our community in Little Buffalo, which is in the boreal forest east of Peace River, about 475 kilometres north of Edmonton.

More than one hundred oil and gas companies were involved. Every year these wells generate about $500 million in revenue for the oil companies and their allies in the Alberta government. To date that adds up to over $13 billion. By 2002, over seventeen hundred oil and gas well sites and countless kilometres of pipelines were situated within our traditional territory. In 2006 alone, the Alberta government approved more than seventy-five new oil and gas wells and seventy-five new pipelines within our traditional territory. All of the companies were issued leases to large parts of Lubicon traditional territory by the Alberta government without any consultation with the Lubicon people.

Roads, oil wells, resource exploitation-related fires, and sharply increased human activity led to a significant decline in our land-based livelihood and reduced us to a state of poverty and dependency. The average number of moose harvested in the community dropped from over two hundred a year before the road was built in 1979 to about nineteen in the mid-1980s. Many traplines were deliberately destroyed. The people working in the oil fields would go out of their way to destroy traps and drive away the wildlife. Before 1979, a Lubicon trapper's average yearly income was about $5,000. Today, it is negligible.

With the onset of resource exploitation has come terrible social and health problems that we never had to face before, such as asthma and other respiratory problems, cancers of all kinds, skin diseases, and miscarriages. In 1985-86, of twenty-one pregnancies there were nineteen stillbirths or miscarriages; in 1987, a tuberculosis epidemic infected a third of the community.

Regulatory Experience

All decisions regarding resource exploitation in our traditional territory are made elsewhere, without our involvement. The benefits of resource exploitation in our traditional territory flow to outsiders. Yet the full impacts of those decisions are felt here, by our people.

Leases and licences to extract oil and gas resources from our traditional territory are issued by the Province of Alberta. The Alberta government claims to have obtained rights over lands and resources in northern Alberta, including unceded Lubicon lands, from the Government of Canada by virtue of the 1930 land transfer agreement. The 1930 land transfer agreement, however, expressly provides that the lands transferred from federal to provincial jurisdiction were transferred "subject to any trusts existing in respect thereof, and *[subject] to any interest other than that of the [Federal] Crown in the same*" (emphasis added).

The Canadian government in turn claims to have obtained rights to these lands and resources in northern Alberta in 1899, including unceded Lubicon lands, through negotiation of Treaty 8 with the original Aboriginal owners. However, the Lubicon people are the original Aboriginal owners of traditional Lubicon territory and, although there have been on-again, off-again discussions about Lubicon land rights between the Canadian government and the Lubicon people going back to 1939 – and effective acknowledgement of Lubicon land rights by both levels of government in Canada – the fact remains that no treaty has yet been negotiated with the Lubicon people, and the Lubicon people have not ceded rights to our traditional territory to anybody in any legally or historically recognized way.

We have never been consulted when leases and licences to extract resources from our territory are sold. No one has ever asked us whether those lands are environmentally sensitive or whether they contain irreplaceable cultural resources or medicines, or are spiritually significant to our people. No one has asked whether our lands and our society can withstand the environmental and social impacts of one oil or gas well, and they have certainly never considered whether we can withstand the impacts of over two thousand wells.

Once the resource companies have bought rights from the province, they decide where and when to begin operations without considering our people's priorities, concerns, or wishes.

Eventually, some companies may contact us to notify us of what they intend to do with our lands and resources. At that point, every major decision regarding the project has already been made, including the most critical decision – whether or not a project should proceed at all. We are asked only to give limited input that may or may not be heeded. The Province does not require resource companies to alter a single thing about their projects should they choose to ignore our input.

All indications are that resource exploitation in our traditional territory is going to expand. Most recently, as traditional oil and gas resources become more difficult to exploit, companies are beginning to move into oil sands exploitation on our lands.

Oil Sands in Lubicon Territory

On 31 August 2004, we learned from newspaper reports that the Province of Alberta had issued oil sands exploration leases to a company with close ties to the Alberta provincial government, called Deep Well Oil and Gas. Deep Well picked up additional adjacent leases in the area through the takeover of another oil company, giving it a majority interest in leases covering over 163 square kilometres right in the heart of our territory.

Deep Well publicly announced that it intended to initiate a large-scale heavy oil extraction program in the unceded traditional Lubicon territory. Deep Well anticipated drilling up to 512 wells in the Lubicon area to extract heavy oil. Heavy oil is the small fraction of oil sands deposits that can flow relatively freely and can therefore be extracted by more conventional methods. Once the heavy oil has been extracted, companies use the surface infrastructure they have built for heavy oil extraction to begin the process of extracting the oil from the oil sands.

The method used to extract oil from oil sands deposits of the type found in our area involves injecting large quantities of steam or superheated water into the ground in order to heat the oil sands, thereby loosening it and making the oil flow. At a time when availability of water for human and agricultural use is becoming a bigger and bigger concern in Alberta, oil sands companies in northern Alberta are not even required to investigate the possibility of using nonpotable groundwater in their tar sands operations. Consequently, the use of potable water in tar sands operations is common in northern Alberta.

Large amounts of potable water – up to four barrels of water for every one barrel of oil extracted – are used in the process of extracting the oil from the oil sands. Even when this water is recycled, at least one barrel of potable water is lost forever for every one barrel of oil extracted. No one knows what impact pumping large amounts of steam into the ground will have on the water table and the fragile boreal forest ecosystem.

In 1979, Shell Oil completed construction of an oil sands pilot project just to the west of our traditional territory. Initially, Shell used water from a nearby freshwater lake for its operations. In short order the level of the lake dropped to the point where the lake froze solid and killed all the fish. Shell subsequently turned to the Peace River to provide the huge volumes of water it requires to operate its small-scale tar sands pilot project.

Imperial Oil built a larger oil sands facility to the east of Lubicon territory near Cold Lake, a large, deep freshwater lake. By 1998, water levels in Cold Lake had dropped to the point where Imperial Oil was no longer allowed by the provincial government to use water from Cold Lake for its heavy oil complex.

Use of groundwater for oil sands extraction can lower the pressure in underground reservoirs. In the Cold Lake area, water pressure in underground reservoirs dropped and farmers have had to drill deeper wells to access well water. Well water in the Cold Lake area has also shown increasing levels of poisonous arsenic contamination and contamination with chlorides and phenols, both of which are associated with heavy oil production.

Another company, BlackRock Ventures Inc., is currently extracting heavy oil on the southwest edge of Lubicon territory. It is not known when BlackRock intends to commence steam operations, but it is clear that steam operations are on the horizon once the company has depleted the more limited and more easily produced heavy oil. BlackRock has recently been bought by Shell Oil – the same company using steam operations to the west of us.

Deep Well Oil and Gas, however, was the first company to propose a major oil sands operation right in the middle of unceded traditional Lubicon territory.

On 1 September 2004, upon hearing of the sale of leases to Deep Well Oil and Gas, we wrote to the company's president and CEO, requesting a meeting between the company's senior executives and the rightful Lubicon owners of the lands and resources around Sawn Lake. Deep Well never replied.

On 2 March 2005, some of our people discovered contractors clearing a large area in our traditional territory on behalf of Deep Well. The contractors were asked to halt operations until a meeting could take place between the company and our chief and council, as requested earlier. Lubicon representatives then made repeated efforts to contact the company, and left numerous messages requesting a meeting.

Still, the company failed to respond.

On 3 March 2005, Lubicon legal counsel received a phone call from Deep Well lawyer Robert Hladun alleging that Lubicon members were blockading the work site and claiming that the blockade was costing Deep Well Oil and Gas "$100,000 a day." Mr. Hladun gave our lawyer a cellphone number at which we could reach a senior Deep Well Oil and Gas official named John Brown. After a number of calls we finally reached Mr. Brown and a meeting was arranged for later that afternoon. Mr. Brown was unavailable to meet earlier because he was meeting with the local RCMP, requesting police assistance in getting access to Lubicon resources.

Mr. Brown claimed to be only a consultant to Deep Well, with no authority to make agreements or commitments. However, he did agree to set up for us a meeting the following week with senior company officials. He promised to phone us back with a date and time.

Mr. Brown failed to phone us back as promised. Deep Well officials continued ignoring repeated efforts on our part to arrange for a meeting.

Through our own research we then learned that Deep Well had not even obtained the required well licences from the Alberta Energy and Utilities Board (EUB). The EUB is the provincial body responsible for issuing licences for oil and gas wells, pipelines, and facilities. We called the EUB to let it know that the company had proceeded to clear-cut approximately eighteen truckloads of spruce trees and twelve loads of poplar without even Alberta regulatory approval.

When the EUB notified Deep Well of this omission, the company began the process to apply for a well licence for a single well at the site. We in turn expressed our concern that assessing an application for a single well was insufficient when the company plans a much larger project on the site. We began to investigate how the larger environmental impact of this project could be assessed.

Environmental Assessment in Alberta

We were told by officials at Alberta Environment that in Alberta no environmental assessment is necessary for a project of this sort until the company proposes to build an in situ plant. This means that unless there is a federal environmental assessment of this project, the companies would be able to drill up to 512 wells, build whatever roads and other surface infrastructure they require, and begin thermal-recovery operations that pump up to four barrels of potable water into the ground for every one barrel of oil they get out – all without any independent assessment of the environmental, social, and economic impacts of these activities on our lands and our people. Instead, we were told that, prior to beginning operations, these companies would be required to obtain approvals from the EUB that would allegedly take into account the company's compliance with Alberta's environmental regulations when granting licences to drill wells.

But the EUB does not assess the cumulative environmental, social, and economic impacts of up to 512 wells and a huge thermal-recovery operation on the lands and people in Lubicon territory. The EUB does not address the question of Aboriginal title. Instead, the EUB will at best assess the company's compliance with Alberta's environmental regulations for each well the company asks to be licensed. Each well will be approved, one at a time, despite the ongoing question of title to the area, and despite the potential effects it may have in conjunction with the other wells, pipelines, and facilities already licensed by the EUB.

By the time the significant adverse effects of a large-scale oil sands operation approved piecemeal by the EUB are readily apparent, the area will be so dramatically altered it will be too late. By the time Alberta Environment is asked to assess the impacts of any proposed in situ plant in this area, the baseline environment will have been forever changed and no meaningful assessment will be possible.

Further, even if the EUB did consider the environmental impacts of a project of this sort, it is unlikely to impose restrictions on oil sands exploitation as a result. In 2005, in a decision approving a hotly contested gas well being drilled directly underneath a lake from which the residents of Bonnyville, Alberta, obtain drinking water, the EUB outlined its philosophy of environmental protection: "Underlying almost all of the written and oral objections [of the residents of Bonnyville] was the view that the residents were not prepared to accept any risk, no matter how small, to the well-being of Moose Lake because of its importance as a source of drinking water and recreational activities. *While the Board acknowledges this sincerely held position, it must weigh the benefits of oil and gas development in the public interest of all Albertans with the potential impacts to the residents and environment in the area*" (emphasis added).

This is why the companies proposing a massive oil sands development at Sawn Lake were so certain they would have no difficulty obtaining approval from the EUB. They began clearing trees on their lease site and moving in a drilling rig before even applying to the EUB for a licence to drill a well in our area. This speaks to the (lack of) seriousness with which these companies treat the EUB approval process and its so-called environmental assessment.

The Lubicon people have serious concerns about the environmental effects of oil sands exploitation. We are concerned that it may contaminate the air, lands, and waters in our territory.

We are concerned that it may damage the fisheries in the nearby lakes – especially given that Deep Well and its partners were given oil sands leases beneath almost all of Sawn Lake and have even proposed drilling wells beneath the lake.

We are concerned about the use of large amounts of potable water for oil sands development. We are concerned that it may deplete the already stressed water systems in our area and damage underground reservoirs. We are concerned that, like at other oil sands developments, toxic contaminants may find their way into underground reservoirs. We are concerned about the unknown impacts of pumping large quantities of steam into the ground in this fragile boreal muskeg ecosystem.

We are concerned about the potential airborne pollutants emitted by the many wells, batteries, and trucks that will litter our lands.

We are concerned about the disposal of contaminated wastes, such as drilling muds.

We are concerned about the effects of so much surface disturbance on the plants, small animals, birds, and large game that are of critical importance for our people and for our culture and way of life. We are concerned about the roads, leases, pipelines, and oilfield camps further fragmenting our forests.

We are concerned about the effect of this large-scale development on our future reserve lands and the Lubicon settlement situated in the area. And we are concerned about the effects unbridled resource exploitation will have on negotiations for a full and just settlement of our rights to these very lands and resources.

Lastly, we are concerned about the effects of increased oil sands exploitation on climate change. Oil sands development in Alberta is expected to add sixty additional megatonnes of CO_2 emissions to Canada's total by 2010. We do not understand how the Canadian government can commit to reducing CO_2 emissions and yet allow developments like this to proceed without ever assessing their impacts on Canada's Kyoto commitments.

Federal Environmental Assessments

Although the Alberta regulatory system requires no comprehensive environmental assessment of projects of this sort, the federal minister of the environment is empowered under Section 48 of the Canadian Environmental Assessment Act to order an environmental assessment of a project if he or she is of the opinion "that the project may cause significant adverse environmental effects on lands . . . in respect of which Indians have interests."

Our people have interests in these lands.

These are lands in which our people have hunted and trapped, gathered medicines and foods, been born, lived, died, and been buried. Our people harvest fish in Haig and Sawn lakes. Our people have had a community at Haig Lake since as long as our Elders can remember. We hope to have a reserve recognized right next to where this proposed development is situated.

These are Lubicon lands. We have never ceded Aboriginal title to these lands and resources in any legally recognized way. The Lubicon Lake Indian

Nation has never signed a treaty with Canada. These lands and resources are the subject of a long-standing land rights dispute between the Lubicon Nation and both levels of Canadian government. For many years now we have been attempting in good faith to negotiate a land rights settlement with Canada and Alberta that would resolve the question of title to this area. Until such time as the question of title is resolved, however, the Lubicon Nation retains unextinguished Aboriginal title to this area.

Even while we attempt to negotiate the issue of land rights in good faith, the Alberta government has sold leases to companies to exploit the very lands and resources that are the subject of negotiations. It has never consulted with the Lubicon Nation about the issuance of oil sands leases on our unceded traditional territory.

Canada, for its part, has a fiduciary duty to the Aboriginal people of the area to protect our interests. Our interests are threatened by large-scale oil sands exploitation in the heart of our unceded traditional territory.

Therefore, we formally requested that the federal minister of the environment, at minimum, have the Canadian government initiate a full-panel review of the project under the Canadian Environmental Assessment Act.

Not surprisingly, the minister did nothing.

Defending Our Lands

When both levels of government have neglected their duty to deal fairly with our Aboriginal land rights; have neglected their legal duty to consult with us before issuing permits, licences, and leases on our lands; and have failed to assess or regulate the environmental impacts created by the companies they have unleashed on us, it is up to us to defend our lands and people as best we can.

On 11 March 2005, our lawyer wrote to Surge Global Energy – which had by this point become a partner of Deep Well but not yet the lead player in the proposed project – explaining Lubicon concerns with the proposed Deep Well project and advising Surge Global Energy of Deep Well's continuing lack of response to our repeated requests for a meeting. Our lawyer received no reply from Surge Global Energy either.

On 30 March 2005, Walter Whitehead, the Lubicon member on whose traditional trapline the project is situated, wrote to then-Deep Well president Steven Gawne advising Mr. Gawne that he'd never been consulted about the proposed project and expressing his concerns about the proposed project. A noted copy of the letter was sent to Surge Global Energy president Frederick Kelly. Neither Mr. Gawne nor Mr. Kelly responded to Walter Whitehead.

Between March and June 2005, we let it be known through the media that the Lubicon people have serious concerns about the proposed project and that we are seeking a meeting with the companies involved. Not one of the involved companies contacted us.

On 8 July 2005, Walter Whitehead received a letter from Surge Global Energy president Frederick Kelly. The letter, dated 22 June, gave Walter Whitehead a "10 day notice as requested of Surge by Alberta Environmental Protection" that Surge has "recently applied to Alberta Environment for access onto the lands covered . . . on the attached survey plan." The letter informed Walter Whitehead that "traps and trap lines affected by this disposition should be removed/relocated at this time to prevent accidental damage."

Walter Whitehead faxed a reply to Mr. Kelly on 11 July telling Mr. Kelly that "it's a little late to warn me about potential damage to traps and trap lines when your contractors already cleared a large site on my trap line without any notification or consultation." Walter Whitehead reminded Mr. Kelly that "the lands you are proposing to tear apart are at the heart of the Lubicon Lake Indian Nation's unceded traditional Territory." With the support of Lubicon chief and council, Walter Whitehead advised Mr. Kelly that "your project will not be allowed to go ahead while you continue to ignore the Lubicon Nation's request to meet and discuss the proposed project." Walter Whitehead received no reply to this fax.

During this time, we enlisted the help of supporters in Canada, Europe, and the United States and began approaching investors in Deep Well Oil and Gas and its new partners Surge Global Energy, Pan Orient Energy, and Paradigm Oil and Gas, and other associated companies. We began telling our story to journalists and stock market analysts. Opposition politicians repeatedly raised the issue in the Alberta Legislature. Environmental groups issued news releases and informed their members. High school students wrote letters to all concerned. We posted photos of the damage done to our land on the internet. And we kept constant watch on the site to make sure that Deep Well and its partners didn't try to move in without our permission.

After months of pressure, the chairman of Surge Global Energy finally made contact with us, in July 2005, and agreed to meet. We met with Surge chairman David Perez on 31 July. At the meeting, Mr. Perez said that his company's share price had dropped from $4 to $1 since the conflict with the Lubicons had begun.

At the meeting, Mr. Perez made a number of commitments. He committed to not using potable or surface water to extract the heavy oil at Sawn Lake. He committed to the companies requiring Lubicon agreement not to oppose their proposed projects before applying to the province for regulatory approval. He committed to treating the Lubicon Nation like another regulatory body.

However, as is often the case, company officials were much more forthcoming with their verbal commitments than with those they were willing to put in writing. Upon trying to document this agreement, we entered into several more months of conflict with the companies and their investors.

Finally, on 14 October 2005, a letter of agreement between Surge Global Energy, its partners, and the Lubicon Nation was signed.

Recognizing that the Lubicon Nation has an unresolved land rights dispute and that the Lubicon Nation opposes resource extraction in their traditional territory prior to resolution of that dispute, the companies agreed to a process that respects the Lubicon Nation's right to stop activities that threaten vital Lubicon wildlife, environmental, cultural, and historical interests.

The procedure established by the letter of agreement is simple. Before applying to the provincial regulatory authorities for approvals, permits, or licences to proceed with any project elements (wells, pipelines, etc.), Surge Global Energy and its partners will present a detailed plan for the proposed activity to the Lubicon Nation and seek Lubicon agreement not to oppose the particular project element. Only after the companies have obtained Lubicon agreement not to oppose the project element will they go forward to the provincial regulatory agencies for permits or licences.

Surge Global Energy also agreed that it and its partners will not use any surface water or potable water in their oil exploitation operations. Further, the companies agreed that no steam or hot-water operations will be conducted without the Lubicon Nation agreeing not to oppose those operations. Damage to the surface lakes and creeks and the general hydrological system in the area from steam or other thermal operations is a key Lubicon concern.

Any drilling sites or pipeline routes will be altered if necessary to address Lubicon wildlife, environmental, cultural, or historical concerns.

Lastly, the companies agreed that any exploitation beyond the initial phase of the project will require consideration of the cumulative environmental effects of the project as a whole, rather than the evaluation of only the effects of each project element in isolation.

Although the letter of agreement represented a positive step toward protecting vital Lubicon wildlife and environmental interests, it remains to be seen how well the agreement is respected in practice. We continue to monitor the implementation of the agreement and ensure that our continuing environmental concerns are respected by the companies.

Ad Hoc Regulation

It is no small effort for us to confront multimillion-dollar oil and gas companies who want to exploit our lands. It requires time, energy, and resources that stretch the capacity of a small community like ours. When our backs are against the wall we are willing to do whatever is necessary to defend our land, our people, and our rights, and through experience we have become very effective at doing so. But ad-hoc regulation of this sort is not a sustainable, just solution to the problems our people face. We need all governments to adhere to their own laws, meet their responsibilities, and resolve the long-standing land rights dispute with our people.

And we are not the only people who believe so.

United Nations Decisions

In 1990, the United Nations Human Rights Committee (UNHRC) concluded in a landmark decision that historical inequities and more recent developments have endangered the way of life and the culture of the Lubicon Cree. The committee ruled that "so long as they continue," these threats are a violation of our fundamental human rights.[1]

Fifteen years later, on 1 November 2005, the UNHRC considered the Lubicon situation for the second time. The UNHRC reaffirmed its earlier conclusion that Canada is violating the Lubicon people's human rights and urged Canada to negotiate a settlement with the Lubicon Lake Indian Nation.

The section of the committee's conclusions on the Lubicon situation states:

> The Committee is concerned that land claim negotiations between the Government of Canada and the Lubicon Lake Band are currently at an impasse. It is also concerned about information that the land of the Band continues to be compromised by logging and full-scale oil and gas extraction, and regrets that the State party (Canada) has not provided information on this specific issue. (articles 1 and 27)
>
> The State party (Canada) should make every effort to resume negotiations with the Lubicon Lake Band, with a view to finding a solution that respects the rights of the Band under the Covenant, as already found by the Committee (in the 1990 Committee decision finding Canada in violation of the International Covenant on Civil and Political Rights over abuse of the human rights and aboriginal land rights of the Lubicon people). It should consult with the Band before granting licences for economic exploitation of the disputed land, and ensure that in no case such exploitation jeopardizes the rights recognized under the Covenant.[2]

In May 2006, the United Nations Committee on Economic, Social and Cultural Rights (UNCESCR), also considered our situation. It concluded that "the Committee strongly recommends that the State party [Canada] resume negotiations with the Lubicon Lake Band, with a view to finding a solution to the claims of the Band that ensures the enjoyment of their rights under the Covenant. The Committee also strongly recommends the State party to conduct effective consultation with the Band prior to the grant of licences for economic purposes in the disputed land, and to ensure that such activities do not jeopardize the rights recognized under the Covenant."[3] Between the time that the United Nations Human Rights Committee ruled that Canadian governments "should consult with the Band before granting licences for economic exploitation of the disputed land, and ensure that in no case such exploitation jeopardizes the rights recognized under the Covenant" and the time the UNCESCR made an identical recommendation, Alberta sold new leases and licences to exploit oil and gas on over sixty-five thousand hectares of our land. The EUB, for its part, approved more than fifty new oil and gas wells and more than fifty pipelines in our territory.

Even while the UNCESCR was considering our case, the Government of Alberta announced an auction for new oil sands leases on over fifty thousand hectares of our lands.

All without any consultation or negotiation with our people.

There is no reason why our people should be excluded from decisions about the future of our lands and resources. All we ask is that we be treated the way any people would like to be treated; with respect for ourselves, our lands, and our rights. Our experience shows us that Canadian governments are not prepared to do so. But our experience also shows us that, with the help of others who share our principles and are willing to stand with us, we can defend our lands and our rights. Until such time as Canadian governments and resource companies are willing to treat us with the respect we deserve, we will continue to do so.

Notes

1 This decision is reported in UN document CCPR/C38/D/167/1984.
2 This decision is reported in UN document CCPR/C/CAN/CO/5.
3 This decision is reported in UN document E/C/12/CAN/CO/5.

7
Population Health, Environmental Justice, and the Distribution of Diseases: Ideas and Practices from Canada

John Eyles

This chapter will locate environmental justice in the discourse of population health, a made-in-Canada set of ideas that now has global reach. Although both discourses (environmental justice and population health) have been vital in extending understanding of the distribution of hazards and the factors that influence both individual and community health, I will argue that they may have implicitly led to a dilution of analysis of root causes of environmental and health inequalities. Examination of environment and health has often been restricted by technical difficulties in dealing with complex relationships, such as measuring environmental impact or attributing a health outcome to particular "causes" given certain mediating factors. Furthermore, there is a disjuncture between much environmental and much health discourse that has led to a marginalization of "environment" as a determinant of health and perhaps of social justice as well. Thus, after outlining the nature of population health, the chapter will explore attempts to see environment as part of health, conceptually and methodologically. It will assert the importance of environment for health and move to examine primarily Canadian examples of this linkage, pointing to their significance in distributional terms. It will conclude by looking at the political and policy challenges of ensuring a resilient environmental health world.

The Discourse of Population Health

In many ways, the appeal to population health is not new. Population health takes us beyond individual health status to consider the overall healthiness of groups and communities. In this way, population health provides a health lens for examining environmental justice. It has been defined as an approach that "aims to improve the health of the entire population and to reduce health inequities among population groups" (PHAC, 2007) by focusing on a range of factors that influence health. Ashton and Seymour (1988) in their essay on the new public health point not only to past practice but also to the quite revolutionary ideas that changed the ways in which health and disease are considered. In terms of past practice, they detail the legacy of nineteenth-century industrialization in northern Europe with its toll of death and disease among the working class living in poor environmental conditions and on low incomes. The public health movement emphasized environmental change to improve air and water quality, food availability, and sanitation. But throughout the twentieth century it seemed that curative approaches – the isolation and treatment of the sick in clinics and hospitals and the development and marketing of therapeutic drugs – had in themselves led to declines in premature mortality and improvements in health status and quality of life. Groundbreaking research by McKeown (1976) showed that declining rates of premature mortality in England and Wales preceded many, if not most, therapeutic advances. This research led to a sea change in ideas, as well as to significant further research in Britain. The Black Report was published in 1980, extending the argument concerning the lesser efficacy of clinical treatments (see Townsend and Davidson, 1982). The report noted marked inequalities in health between the social classes and that a class (as measured by occupation) gradient existed for most causes of death. Thus not only were social and economic forces implicated in health but also in its distribution between groups. This finding has led to a research industry documenting health and illness gradients. The definition of health was among the changing ideas. As the report presciently noted in its opening chapter, "there are different meanings of 'health.' On the one hand, 'health' can be conceived in the outcome of freeing man from disease or disorder, as identified throughout the history of medicine. On the other hand, it can be conceived as man's vigorous, creative and even joyous involvement in environment and community, of which presence or absence of disease is only a part" (Townsend and Davidson, 1982, pp. 49-50). In this there is the call to attend to population health or the social determinants of health, that is, those forces – income, environment, gender, and culture – that shape health status.

A similar discourse was also being created in Canada. The Lalonde Report identified human biology, environment, lifestyle, and health care organization as determinants of the public's health, perhaps the first occasion on which health was seen as being produced by more than health care in this country (Canada, 1974). This was followed up by a later federal document – the Epp Report (Canada, 1986) – a contribution to the World Health Organization's (WHO) *Achieving Health for All* strategy. As Raphael (2003) has noted, this report emphasized influencing the social determinants of health as a means of reducing the health inequalities between income groups. Furthermore, these mid-1980s debates led to the formulation of the Ottawa Charter for Health Promotion (WHO, 1986). This charter identified peace, income, education, shelter, sustainability, food, equality, and social justice as prerequisites for health but essentially advocated strategies for individuals and communities to prevent disease and promote health, largely through healthy living.

Canadian academic discourse also engaged in the population health debate. The Population Health program of the think tank Canadian Institute of Advanced Research put forward a conceptual framework for thinking about why some people are healthy and others not (see Evans and Stoddart, 1990; Evans, Barer, and Marmor, 1994; Frank, 1995). Emphasized are not medical inputs and utilization but social, cultural, and economic factors. Specifically, in a population, a high level of economic development and a relatively equitable distribution of wealth are associated with higher levels of health status, whereas at the individual level immediate social and economic environment and its interactions with psychological resources and coping skills are key (see Frank, 1995). Furthermore, a suite of determinants is put forward: income, social status, social support networks, education, employment and working conditions, physical and social environments, biology and genetic endowment, personal health practices and coping skills, healthy childhood development, and health services (see Evans et al., 1994). This suite of determinants has been codified by Health Canada (1999). Health Canada has also collated the evidence base on these forces (Health Canada, 2003). A particular focus has been the health of Aboriginal peoples (Allec, 2005; Eyles, Birch, and Chambers, 1994; Health Canada, 2000).

The development of these "determinants" is international, with specific emphasis on the social ones. Thus, WHO has listed the social gradient ("class" in the Black Report), with stress, early life, social exclusion, work, unemployment, social support, addiction, food, transportation as its determinants (Wilkinson and Marmot, 2003). The US Centers for Disease Control and Prevention (2005) puts forward social-economic status, transport, access to services, housing, discrimination by social group (class, gender, race) and social and environmental stressors. Working from Canada, Raphael (2004) puts forward eleven determinants: Aboriginal status, early life, education, employment and working conditions, food security, health care services, housing, income and income inequality, social safety net, social exclusion, and unemployment and employment security. Internationally, the Commission on the Social Determinants of Health, created by WHO, is now in place to develop knowledge networks around themes – early child development, employment conditions, globalization, determinants of women's health, the health system as a social determinant, urban settings, social exclusion as measurement issues. A further network will look at disease control programs for infectious epidemics (see Irwin et al., 2006).

What should we make of these lists of the determinants of population health? We can note the variations in language used to describe specific factors; the removal of biology and genetics from consideration, although the biosocial and psychological dimensions remain as research frontiers in medicine and psychology; the downplaying of health care interventions; and the specificity of certain realities – addiction, food security, and disease control are examples. But overall, the population health discourse has advanced our understanding of health. Through it, blaming victims of poor health for "poor" choices has largely disappeared. Raphael (2003) notes the lack of evidence of lifestyle choices being a threat to and salvation of health. And WHO comments, "The greatest share of health problems is attributable to the social conditions in which people live and work" (Irwin et al., 2006, p. 749). Yet for our purposes, there is a significant silence. Population heath discourse may well do justice to the complexities in the production of health, but the term *environmental* becomes so broad as to be meaningless or so limited (e.g., to refer to transport, employment security) as to apply only to specific populations. We will now explore why there has been silence on the physical environment.

Putting Environment into Population Health Discourse

Is environment the forgotten determinant of health? And if it is, why is this so and with what effect? To foreshadow our argument, it is not so much forgotten as existing in its own discourse universe, which may mean that for some debates it is marginalized. Certainly, at high levels of discourse, physical environment is discussed. Thus, Health Canada (2001) notes that "the physical environment is an important determinant of health." But it goes on: "At certain levels of exposure, contaminants in our air, water, food and soil can cause a variety of adverse health effects, including cancer, birth defects, respiratory illness and gastrointestinal ailments. In the built environment, factors related to housing, indoor air quality and the design of communities and transportation systems can significantly influence our physical and psychological well-being." From this, we may note numerous environmental factors associated with many possible adverse health outcomes, the associations of which are likely to be mediated by the other social and biological determinants of health discussed in the previous section. Indeed, this identified complexity of the issue has been noted by several authorities that have entered the debate. For example, the Institute of Medicine in its reports on environmental justice (1999) and environmental health indicators (2004) emphasizes communities of concern, suspected as having disproportionately high levels of exposure to various environmental stressors (toxics, chemicals, biologics, and allergens, as well as light, noise, odours, and particulate matter), but is left appealing for great commitment to a coordination of surveillance to drive ameliorative actions. Furthermore, WHO (1999, 2002) adds other phenomena, which it terms "modern hazards" and which are often associated with unsustainable development practices and including water pollution from populated areas, industry and intensive agriculture; urban air pollution from vehicular traffic, coal power stations, and industry; climate change; stratospheric ozone depletion; and transboundary pollution to the possible environmental pathways of exposure that may lead to adverse outcomes, complementing traditional hazards such as poverty, lack of access to safe drinking water, inadequate basic sanitation in the household and community, indoor air pollution from cooking and using biomass fuel, and inadequate solid waste disposal (see also Eyles and Consitt, 2004). It is perhaps no surprise that researchers concentrate on one type of exposure at a time, but even then, given confounding and/or mediating factors, it is often difficult to establish a scientifically robust association between the exposure and the outcome.

But such science is necessary. As Briggs, Corvalán, and Nurminen (1996) note, "Human exposure to pollutants in air, water, soil and food – whether in the form of short-term, high-level episodes or longer-term for level exposures – is a major contributor to increased morbidity and mortality. The disease burden attributable to these exposures is not known with any certainty, however, because levels of general environmental pollution fluctuate greatly, methods for analyzing the relationships are incompletely developed and the quality of available data is generally poor" (pp. 1-2). But it is the fluctuations even in background levels that make investigating the relationships between physical environment and health so vital. The word *fluctuation* implies an uneven distribution of pollution, which may further imply the existence of exposure and health inequalities. And if that most valued of outcomes – good health – is unevenly distributed, then investigation and action are not only worthwhile but vital. This requirement has been discussed in both the environmental justice and population health literatures. For example, Buzzelli (2007) notes that the examination of the disproportionate distribution of health hazards has been stalled despite twenty years of research because of these persistent methodological and conceptual problems. In the United States, the relationship between residential location and segregation and health disparity is well understood (see Williams and Collins, 2001), but as Brulle and Pellow (2006) note, the role of exposures to toxic pollution on community health is still sorely lacking.

There have, however, been important conceptual advances advocating a nuanced discourse on these relationships. Northridge, Stover, Rosenthal, and Sherard (2003) explicitly see the population health paradigm as a way forward in assessing the distribution of environmental hazards, with race and gender being implicated in the effects of particulate matter (PM_{10}) on overall mortality (see Zanobetti and Schwartz, 2000). Evans and

Kantrowitz (2002) suggest that socio-economic status (SES) may well be associated with environmental quality and that in turn may affect health; that is, the relationship between SES and health may be affected by environmental exposures. But Evans and Kantrowitz (2002) add: "This is not equivalent, however, to the conclusion that SES effects on health are caused by differential exposure to environmental quality . . . But [rather] that the SES health link is mediated by environmental quality" (pp. 303-4). This local level discourse on the importance of environment for population health is furthered by Frohlich, Corin, and Potvin (2001) with the development of "collective lifestyle" as an expression of a shared way of relating and acting in a given environment. In other words, social inequalities in diseases at the local level are shaped by exposures and resources but also by capabilities and social practices. Buzzelli (2007) uses this approach to bring together population health and air-pollution epidemiology that does perhaps remain similar to earlier calls to integrate population health and environmental epidemiology (see Eyles, 1999).

Can such a discourse be made empirical? Yes, but other dimensions of the discourse must be outlined. Population health and environmental epidemiology are necessary but not sufficient elements of a conceptual framework. Smith and Ezzati (2005) suggest that a risk transition framework is an improved heuristic. This is similar to WHO and its identification of traditional and modern hazards with associated disease states (namely infectious and nutritional) with the traditional and its identification of cancer and cardiovascular, chronic, and degenerative diseases with the modern. But the strengths of Smith and Ezzati's framework (2005) derives from their making it overtly political: "To put it starkly [, we note] a tendency for societies to sweep environmental health problems out of the house and into the community during the first stages of development and then out from the community to the general global environment during the later stages" (p. 295). In other words, environmental health risks are now communal and shared. This is close to Beck's argument (1992) for the democratization of risk. Yet risk outcomes, especially poor health, remain stubbornly localized and maldistributed. Smith and Ezzati (2005) recognize this by pointing out the continuing existence in developed societies of community-based exposures (e.g., urban ambient air pollution, lead pollution, occupational risk, traffic accidents). Furthermore, there is risk overlap in that risk factors coalesce, most obviously seen in the urban slums of poorer countries. And new types of risk can be generated such as personal care products in waste and groundwater. Risks can also be transferred through control attempts, for example, pesticide use for mosquito control for West Nile virus in the United States and Canada. Finally, there can also be synergistic effects, magnifying vulnerability.

Seeing exposures as risks that may increase the vulnerability of populations differentially is an important lens for the environment-population health discourse. Vulnerability is not merely biophysical but also socially amplified or minimized by context (see Turner et al., 2003; Clark, Elliott, and Eyles, 2006). Brown (1995) highlights how such vulnerabilities are produced and emphasized not only in the location of hazards and the role of regulatory agencies but also in the story of industrial change, land-use, and social history, and as such shifts in the composition of communities. In other words, putting environment into population health requires a full social analysis of the context in question. In this an important ingredient must be the temporal dimension. As Hertzman, Frank, and Evans (1994) put it, "Time lays another serious trap in the way of understanding the determinants of health" (p. 82). They point to four ways in which time may affect a population's health and specifically the association between an event (exposure) and outcome (see also Murray, Ezzati, Lopez, Rodgers, and Vander Hoorn, 2003). First, many health effects are delayed: there is an incubation period or an *elapsed time* in the effect. Second, *biological time* is embedded in the life cycle of individuals and may create windows of vulnerability. There has, for example, been much research on reproductive health and the window from conception to birth. Third, *cumulative time* points to the need for toxicants to accumulate before disease emerges. This is particularly important in occupational and possibly residential settings (e.g., silicosis and trihalomethanes in drinking water respectively). Finally, Hertzman et al. (2004) point to *historical time* in that conditions may be ripe for certain exposures for whole populations (e.g., infectious outbreaks such as SARS and influenza, exposure to radiation around weapons-producing facilities). Historical time – the course of history and why things happen to particular groups in particular places – highlights the centrality of social analysis for environment and population health. Thus, exposures and risks may be disproportionally allocated and distributed because of the historical production of societal position.

Environmental justice itself encompasses such a conception of time and its societal implications. But a progressive discourse may look to the future as well as to the past. Taylor (2000) has noted the linking of environment, race, class, gender, and social justice in the environmental justice paradigm. Agyeman and Evans (2004) argue that just sustainability focuses on quality of life, justice, and equity in resource allocation and on living within ecological limits. For understanding health, this conception adds a historical time (intergenerational) and equity focus. But to link health outcomes explicitly to just sustainability adds emphasis to quality of life such that it emphasizes individual and community health, providing significant outcomes (reduction in premature mortality and disability) to resource allocation and sustainability decisions. Population health is then key for environmental justice, and just sustainability and justice to the population health. Analytically, the developing lens of political ecology captures much of this broad discourse, as seen in studies of contamination generally (Eyles, 2001) or of Sydney tar ponds toxins on reproductive health (Haalboom, Elliott, Eyles, and Muggah, 2006). This lens encompasses the historical development of environmental conditions within a political context and within these examples there is a focus on health and there may in fact be an important task – beyond the scope of this chapter – to look at the relationships between environmental justice, just sustainability, and political ecology.

A Short Interlude on Method

How might we investigate these relationships? Does it depend on bringing together population health and environmental epidemiology? And is this bringing together possible? In fact, it may not be, largely because there is tremendous variation in both the temporal and the spatial scales of problems to be investigated. If this is the case, then it is likely that methods directed to different scales are needed. Thus, as in any research, the research question must determine the method used. In examining the associations between specific exposures (e.g., particulate matter, lead, PCBs, flame retardants) and particular health outcomes (e.g., lung function, intellectual impairment, cancer), conventional epidemiologic methods are appropriate and can indicate the relative or elevated risk of a health outcome from an exposure. Furthermore, great interest has been shown by epidemiology in trying to assess susceptibility to disease (see Vineis and Kriebel, 2006). This interest may well assist in understanding individual health; at the population level, there is additional uncertainty in examining the relationships and interactions between individuals, groups, exposures, and outcomes.

Although useful, conventional methods are not capable of completely assessing population-level risks across space and time in a just sustainability framework. As Dunn (2006) notes, population health must explain the differential distribution of health status, while accounting "for a complex background of both slowly evolving and punctuated historical changes and an increasingly complex set of geopolitical relations across the globe" (p. 572). It is thus a framework that provides a conceptualization of structure of a system (the production of health) and its interaction with other systems (e.g., the environment, regulatory agencies). To investigate these interactions at the population or community level requires a method that can assess health in relation to these systems, namely a modified form of (environmental) health impact assessment (HIA). WHO (2005) notes that HIA provides decision makers with information about how any policy, program, or project may affect the health of people. But it is possible to modify HIA and use it analytically in terms of assessing the nature of population health status and the antecedents of this state, especially as HIA's guiding principles are democracy (allowing those affected to participate), equity (ensuring the discovery and knowledge of vulnerability), sustainable development (considering intergenerational issues), and the ethical use of evidence (using quantitative and qualitative evidence in open, transparent ways). Such assessments seemed to hold great promise for assessing impacts (in outcome exposures) on a population (see Scott-Samuel, 1998), but concern has been raised over their rigour, leading Parry and Stevens (2001) to suggest that different levels of assessment could be undertaken. Their full assessment, for example, involves extensive systematic literature review, secondary analysis of existing data, collection of new data (quantitative and qualitative), extensive quantification and sensitivity analysis, and full participation of stakeholders to tell the political ecological story of exposure-outcomes relations. The investigation of the Woburn, Massachusetts, "leukemia cluster" is illustrative with its utilization of biogeographical and hydrological surveys, disease registries, interviews, and stakeholder involvement through "popular epidemiology" (Brown, 1992; Fuchs, 1996; Harr, 1996). Clearly, the political ecology of environmental health requires combined methods to discern the social complexity of health inequities and their trajectories in time and space.

Selected Geographies of Population Health and Environmental Justice in Canada

To examine geographies of population health in an environmental justice framework in Canada, studies and commentaries that have used a variety of methods to tease out the exposure-outcome relationship will be employed. To do this, linked examples will be used. First, temporal scales will be illustrated with respect to environmental exposures and reproductive health effects, highlighting generations at risk and intergenerational exposures. A record of these exposures can be seen in the body burden of contaminants: How are these exposures and burdens distributed? One way is adverse reproductive outcomes; their potential environmental associations have been a focus of attention for more than a decade (see Colburn, Dumanoski, and Peterson-Meyers, 1996). As Hollander (1997) notes, "Reproductive health is exquisitely sensitive to characteristics in an individual's environment, including physical, biological, behavioural, cultural and socioeconomic factors" (p. 82). Such health may be affected by the differential distribution of a series of health hazards. Malnutrition can increase a pregnant woman's susceptibility to poor outcomes. Women prescribed DES, a synthetic estrogen, to prevent miscarriage from the late 1940s to the early 1970s had a small but significantly elevated risk of breast cancer and their daughters developing vaginal clear-cell adenocarcinoma. Exposure to lead is associated with fertility impairments in both men and women and the risk of spontaneous abortion and stillbirth. Several solvents are associated with birth defects and spontaneous abortion risk, while pesticides have been linked to intrauterine growth retardation as well as chromosomal defects and malformations (see Hollander, 1997; Schettler, Solomon, Valenti, and Huddle, 2000; OCFP, 2004). Further research is required on the distribution of these insults among particular populations and whether they would add significantly to the burden of illness, that is, whether environmental exposure is significant given other contributions. It may be noted that certain occupational groups and agricultural and rural regions may be adversely impacted, while those communities dependent on specific industries – particularly oil-based and chemical – may add such problems to other poor outcomes. For example, the Chippewa community near Sarnia, Ontario, has seen more girls than boys born over the past ten to fifteen years (see Mackenzie, Lockridge, and Keith, 2005), a possible indicator of environmental insult.

Much interest in reproductive health has emphasized hormone-mimicking or endocrine-disrupting chemicals. The endocrine system produces a variety of hormones, which trigger specific biochemical responses (Hollander, 1997; Schettler et al., 2000). Several synthetic compounds can interfere with hormonal activity. Some mimic natural hormones and may strengthen response. Others bind to receptors, often triggering a weakened response. The best-known human-activity-produced disruptors are birth control pills and DES, dioxins from waste incineration and industrial processes, and PCBs once widely used in electrical equipment and adhesives. Alkyphenols and phthalates as well as pesticides have also been implicated. For women, the most important effect is increased risk of breast cancer in which DDE, a breakdown product of DDT, is implicated. In a study of Toronto and Kingston women, Wolcott et al. (2001) show the DDE exposure was significant for the outcome, although the differences between the cities have not been explored. They do report that the Kingston sample had larger and more well-differentiated tumours. For men, falling sperm counts, undescended testicles, and hypospadias have been documented. Younglai, Collins, and Foster (1998) note declining semen quality among Canadian men and Paulozzi (1999) shows an increased incidence of hypospadias. Regional differences in sperm quality have been found in several studies, but differential environmental exposure is only one of the several possible explanations (e.g., on the sample of men itself, methodological issues, see Jouannet, Wang, Eustache, Kold-Jensen, and Auger, 2001; also Soloman and Schettler, 2000).

Associated with reproductive effects and the transmission of exposures through reproduction are studies that have focused on the body burden of contaminants, a focus that emphasizes the intergenerational scale of pollution and health and justice. These studies are beginning to reveal that our bodies harbour pesticides, flame retardants, lead, nicotine-metabolites, and other toxic human-made chemicals. For example, women in San Francisco have three to ten times the levels of flame retardants in their breast tissue than European or Japanese women (Petreas et al., 2003). Similarly, women and infants in Indiana and California had twenty times the level of flame retardant present in women of Sweden or Norway (Schecter et al., 2003). In fact, in a review, Solomon and Weiss (2002) note that, globally, levels of organochlorine pesticides, PCB, and dioxins in breast milk have declined over the past few decades, while the levels of PBDEs are rising. However, data exist only on a limited range of contaminants.

Nongovernmental organizations have tried to extend this coverage. In the United States, the Environmental Working Group (EWG, 2005) listed the presence of 261 chemicals in umbilical cord blood, many of which are carcinogenic, are toxic to the brain and nervous system, or have caused birth defects in animal tests. By far the greatest number of chemicals found was polychlorinated and polybrominated chemicals, the former from lubricants, oils, varnishes, and incineration, the latter from flame retardants. Newborns are therefore products of industrial society almost in a literal sense. It should be noted that cord blood samples from only ten children were used. In a later report, EWG (2006) extended this work to look at pollution across the generations: in other words, the intergenerational inheritance of pollution. Again, the sample size was small. The blood or urine of eight women was examined and found contaminated with an average of thirty-five consumer product ingredients, including flame retardants, plasticizers, and stain-proof coatings. These substances bioaccumulate in the mothers but are also found in their daughters, remaining for varying periods. According to the Environmental Working Group (2006), "the estimated age by which a daughter will purge 99 percent of the inherited pollution ranges from one day for phthalate plasticizers to one year for mercury, to between twelve and sixty years for common flame retardants and stain-proofing chemicals, to longer than a lifetime, 166 years, for lead."

Environmental Defence carried out similar work in Canada, examining pollution levels in both children and adults. In a survey of families, it detected 46 of 68 chemicals listed for six adults and seven children (Environmental Defence, 2006). Similar outcomes to those found in the US subjects were present. For the perfluorinated chemicals (PFOA, PFOS, and PCH_xS) levels were higher in children than adults. Of the substances detected, 38 were carcinogens, 23 hormone disruptors, 12 respiratory toxins, 38 reproductive or developmental toxins, and 19 neurotoxins. In its adult study, Environmental Defence again used volunteers: eleven people from across Canada (Environmental Defence, 2005b). Importantly, it noted the paucity of reliable tests to detect many chemicals and the high cost of analyzing human samples. Environmental Defence limited its analysis, therefore, to 88 substances. Sixty of these were detected and these are similar to those found in other studies. The report cautiously notes that PFOS contamination is likely widespread in the Canadian population, but pollution does not remain at its point source, and that the north – one volunteer – may have the highest level of mercury, as well as PCBs and organochlorine pesticides.

There is then in this story of intergenerational pollution (and injustice) a link of geography. We now turn to the second element in the linked case studies: exposure at the regional level. Environmental Defence (2005a) has examined pollution in the Great Lakes basin. Using data from the National Pollution Registry, it noted the increase in the release of core chemicals from virtually all sources between 1995 and 2002. In fact, most releases are to air and these have increased by 46 percent. The greatest significant increase is to water (382 percent). The greatest release to air in 2002 was in the Lake Erie basin, mainly because of power generation in Nanticoke. This basin also saw the largest release to landfill, mainly from a hazardous-waste management site in Corunna. Facilities in the Lake Ontario basin have the largest releases to water (mainly from sewage treatment) and pollutants transferred off site to landfills. This does not mean that the other parts of the Great Lakes basin are unaffected. For a relatively small number of facilities, the Lake Huron basin (mining and smelting) and the St. Lawrence River (industrial and power production) have large releases, especially to air. But this regional geography of pollution requires a finer grained analysis to link it to population health or environmental justice. This has been done in part by Elliott, Eyles, and DeLuca (2001) in their study of health outcomes in the Canadian areas of concern, pointing to the coincidence of problems in the Detroit, Niagara, and St. Clair rivers. But the linking of health risk to hazardscape (see Cutter, 2001) is methodologically challenging.

These linkages may, however, be clearer in the Arctic, where there is the additional consideration of Native populations (see Trainor et al., this volume). Environmental change in the Arctic has been a concern for decades; this concern is now given added impetus by the security implications of an ice-free passage in the Arctic (see Polar Research Board-Marine Research Board, 2006). The warming of the Arctic is consistent with climate change and its amplification through feedbacks such as snow-ice radiation (Overpeck et al., 1997; Wang and Key, 2003). This apparent, fundamental change is not the only environmental insult. As Furgal and Keith (1998) note, the Arctic was once considered a pristine environment. There are few local human-produced sources of pollution. There do exist within the Canadian Arctic abandoned radar sites and military installations, mining, and garbage dumps. But most Arctic contamination is the result of transport from industrial and agricultural sources in the mid-latitudes, raising questions of regional environmental equity, especially for pollutants such as pesticides, PCBs, and PBDEs (see Suk et al., 2004). Such concerns are not, of course, limited to the Canadian Arctic (e.g., see Poikolanian, Tombre, and Kallenborn, 2004).

What is the population health impact of these environmental changes? This is not an easy question to answer, as any health effects from such sources are not the only effects present. Lifestyle and economic changes have had significant influence on the growth of chronic diseases (cardiovascular, substance abuse, diabetes) in these populations (see Bjerregaard, Young, Dewailly, and Ebbesson, 2004). Furthermore, as Dewailly and Weike (2003) note, environmental epidemiologic studies are difficult to carry out in the Arctic because of low population numbers, remoteness, cultural context, and social and behavioural factors. They also note that the mixing of contaminants may be different in the Arctic than at lower latitudes, meaning that exposure profiles may differ and not be transferable from other studies. But as Suk et al. (2004) note, pollution reaching the Arctic is usually too diluted to be a major human hazard, although its bioaccumulation and biomagnification in plants and animals and then through food chains and diet to humans are important. Local or traditional foods – country foods – are vital not only for nutrition but also for cultural and spiritual reasons, significant in the Native way of life. These become a significant exposure pathway. Studies analyzing material and umbilical cord blood have found higher levels of organochlorines, mercury, lead, and cadmium in Artic as opposed by non-Arctic populations (Butler-Walker et al., 2003; Butler-Walker, Houseman, and Seddon, 2006; Tsuji et al., 2005), although smoking rates and local pollution sources make attribution to transboundary pollution or diet difficult. But Dewailly and Weike (2003) note an association between higher levels of some pollutants and consumption of marine mammals in particular. High levels of PCBs and mercury contamination are found among the Greenland Inuit and are also present in the Baffin and Nunavik regions in Canada. But country food consumption can itself add to health concerns, as lead levels are largely linked to the use of lead shot in hunting. The Arctic thus presents a complex picture. Global-scale events (e.g., climate change, transboundary pollution), leading to regional inequities (e.g., the migration of contaminants from source regions), combined with culture and local practices (e.g., hunting with lead shot, lifestyle decisions), still mean a traditional way of life and the health of local populations are at risk.

The final example of population health and environmental justice is at the local level and in some ways builds on the earlier picture of the regional geography of pollution. For this we will examine air quality and health. Clean air is seen as a basic need for health and well-being. Air quality has been compromised for centuries with the burning of fossil fuels and particularly with industrialization. It took major adverse health events in the 1940s and 1950s in Europe and the United States for scientific and policy attention to become focused on the problem. As the same factors appear to be at work, poor air quality and climate change are regarded as having similar consequences (see Suzuki Foundation, 1998). And their consequences seem severe. Burnett, Camak, and Brook (1998), in a study of eleven Canadian cities, calculated that almost 8 percent of all nontraumatic mortality is attributable to air pollution.

But the picture here is also a complex one. Not only are there different types of pollutants with different health effects but their distribution varies between cities and parts of cities. First, ambient particles, especially those of small diameter, have many impacts on respiratory, cardiopulmonary, and cardiac health, as well as on hospital admissions (Delfino, Murphy-Moulton, Burnett, Brook, and Becklake, 1997; Katsouyanni, 2003). Ambient particulate matter is a product of combustion. Second, ozone (O_3) is produced by the action of solar radiation on other pollutants such as volatile organic compounds (VOCs) and nitrous oxides (N_{ox}). Ozone affects lung function and airway responsiveness. Goldberg et al. (2001) found a 3.3 percent increase in daily deaths in Montreal in the warm seasons, linked to ozone levels. Third, nitrogen dioxide is a result of vehicle emissions with similar effects as ozone. In a study of traffic air pollution and mortality in Hamilton, Finkelstein, Jerrett, and Sears (2004) note an advance of mortality rate by two and a half years associated with residence near a major road. Fourthly, sulphur dioxide SO_2 derived from fossil fuel combustion has similar health effects as particulate matter (Katsouyanni, 2003). Similarly, Villeneuve et al. (2003) observed an increased mortality in risk from SO_2 exposure in Vancouver, especially for those with lower socio-economic status. Finally, carbon monoxide is identified as yet another problem. Carbon monoxide results from incomplete combustion of certain fuels and is more likely found in indoor than outdoor environments. This dizzying level of detail is included here to show the difficulties of associating specific exposures and hazards with health outcomes, and developing policy to address environmentally produced health outcomes and inequities.

The distribution of these sources and their effects is not uniform across Canada. Burnett et al. (1998) note that increased risk of death from nitrogen oxide concentration is greater in London, Calgary, and Vancouver; for ozone, it is higher in Quebec, Montreal, and Hamilton. For all measured pollutants, the increased risk in mortality varied from 3.6 percent (Edmonton, Windsor) to 11 percent (Quebec), with Hamilton and London also showing increases of over 10 percent. Advances in scientific measurement have tended to result in these risks increasing in size. In 2000, for example, the Ontario Medical Association (2000) estimated that air pollution contributed to 1,900 premature deaths in Ontario. In 2005, this figure had been raised to 5,800 to reflect the impact of a greater number of pollutants and the cumulative and long-term effects of smog (OMA, 2005). Similarly hospital admissions have increased from 9,800 (OMA, 2000) to 16,800 (OMA, 2005) and emergency room visit from 45,250 to 59,700. OMA has also calculated the economic costs of air pollution to include productivity, health care costs, pain and suffering, and loss of life, as well as the regional impact of poor air quality. For example, it has estimated that in 2005, adverse air quality resulted in 290 premature deaths in Hamilton, a similar number to Ottawa, which has twice the population.

Differential health impacts are therefore related to the distribution of environment conditions *between* cities, resulting in regional disparities of costs. Other research has examined what is happening *within* cities. O'Neill et al. (2003) note the complex pathways between exposure, neighbourhood, socio-economic position, and health outcome with their review of (mainly US) studies pointing to more adverse outcomes among those in lower socio-economic positions. Much work on intra-urban differences has been carried out in Hamilton. Finkelstein et al. (2003) show that mean pollutant levels tend to be higher in lower-income neighbourhoods. Furthermore, income and pollution levels were associated with mortality differences. Those with lower incomes and higher particulate levels had a 2.6 times greater risk of death from nonaccidental causes than from all other causes. Interestingly, the lower income–lower particulate group was 1.8 times more likely, and those with higher incomes–higher particulate matter 1.3 times more likely to die. Despite sample and measurement issues, this study shows an environmental as well as income effect on premature mortality. Intra-urban differences have been found in Hamilton using census and total suspended particles (TSP) data. Jerrett, Buzzelli, Burnett, and DeLuca (2005) show a relative risk of premature mortality for TSP exposure of 1.2 for women and 1.3 for men. Associations with cardiovascular and cancer mortality were largely significant. The health effects of exposure to particulate matter were not eliminated by including socio-economic, demographic, or lifestyle factors. Jerrett et al. conclude that adverse health effects vary by area in Hamilton. The area effects could, however, be modified by some socio-economic variables such as low educational attainment and high manufacturing employment (see Jerrett et al., 2004). Other dimensions of the population or exposure may further modify the relationship. For example, Buzzelli and Jerrett (2004) note the proportion of Latin-Americans in a census tract is positively associated with pollution exposure, whereas Asian-Canadians are negatively associated. And Keller-Olaman et al. (2005) add indoor exposure to the analysis (albeit based on self-report) and note the complex relationships between indoor, pesticide, and work exposure with a variety of occupational, neighbourhood, and socio-demographic variables. Thus, while environment is a key variable for understanding the geography of population health, definition of exposure and outcome and scale of analysis and the inclusion of mediating factors will affect that geography.

Conclusion

The different geographies of population health point to the complexities of scale in establishing meaningful health outcomes, a task made more challenging by considerations of environmental justice and just sustainability. The three examples – embodied burdens intergenerationally associated, inequities produced by regional variation, and local disparities – are made complex by their historicity and interweaving of exposure-pathway-mediating factors-outcomes. On a more practical level, the Arctic and air-quality issues are scenes of jurisdictional conflict, internationally and federally-provincially, respectively. Although both concern the federal government's responsibilities for security and foreign relations, health and environment matters fall under provincial purview. The agendas of mitigating damage in the Arctic caused by transborder production and transport and reducing localized sources of air pollution involve federal, provincial, and municipal expenditures. Examination of any budget from any level of government demonstrates little financial attention is paid to the environment or environmental impacts on health. In Ontario, for example, the health care budget dwarfs all others and its public health share (where environmental health expenditures would be primarily located) is a tiny portion at around 2 percent of total budget. The Ontario health budget is approximately $35 billion for 2006-2007. By comparison, the Ontario Ministry of the Environment is allocated $302 million, Natural Resources $682 million, and Northern Development and Mines $347 million. Thus, health (largely doctors, hospitals, and drugs) consumes 41 percent of the provincial budget, while the broadly defined environment ministries comprise just over 1 percent (Ontario, 2006). To this small provincial contribution, federal environmental initiatives have been largely shelved until 2010. A fund to help cities – largely with infrastructure – has been established (Canada, 2006). The municipalities themselves have been left struggling, especially in Ontario, with the downloading of welfare costs by the province, which has largely squeezed out or reduced other expenditures. This is unfortunate, as it has been shown that pollution is associated with higher health care expenditures, while per capita municipal environmental expenditures are associated with lower ones (see Jerrett, Eyles, Dufournaud, and Birch, 2003).

So environmental expenditures matter and are associated with demand for health care as measured by expenditures. How might these figure into a range of policy options to improve population health and enhance environmental justice? Certainly, the contribution of environmental degradation (and improvement) to the health of the population has to be recognized (see also Litt, Tran, and Burke, 2002). Given the complexities of scale, a range of policy options addressing areas and individuals and groups is needed. Area policies may be useful for some dimensions of population health-environmental justice, as they can positively discriminate in favour of the disadvantaged. But as Thompson, Atkinson, Petticrew, and Kearns' review (2006) of urban regeneration programs in the United Kingdom suggests, area policies have to be crafted carefully to ensure these positive outcomes (and to minimize negative, often unintended consequences). Individual-, family-directed policies are necessary to ensure access to resources for health and a safe environment either through income guarantees or direct service provisions. But again, conflicting jurisdictional responsibility may overwhelm policy initiatives. It is not only resources that are necessary; so too is a commitment to a specific kind of social order. This is more than paying lip service to the welfare of others. It is a recognition of the fundamental inequities in health that derive from socio-economic position (see Regidor, 2006) and of the overriding but challenging value that health equity matters in practice, not just principle (see Lindbladh, Lyttkens, Hanson, and Ostergren, 1998). The arguments in this essay show the lie of the "democratization of risk and hazard in modern society." The impacts of many environmental insults fall disproportionally on some more than others because these unequally distributed insults are determined by class, culture, and locality. Furthermore, children are at particular risk. They are products of industrial society, its test sites for contaminants from products or from environmental degradation. Children's health and that of disadvantaged populations are powerful and important tools for seeking environmental change (see Burger, 1990), but the evidence is not easily gathered because of the complexity of the hazard exposure–health outcome relationship and the downplaying of environment in population health discourse. Difficulty does not mean impossibility. These linkages must be advocated and researched. So, too, must the full social, political, and historical context in which environmental insults and adverse health outcomes are occurring.

Canadians have contributed to the research, policy, and discourse on the social determinants of health. Inequity results in decreased population health. Any just civil society must value health equity. The right of citizenship and ethical deliberation demand such attention. And without this value-based commitment, we will never stay the course to achieve equitable health outcomes or environmental justice across the generations in Canada.

References

Agyeman, J., and B. Evans. 2004. Just Sustainability. *Geographical Journal, 170,* 155-64.

Allec, R. 2005. *First Nations and Wellness in Manitoba.* Winnipeg: Manitoba Health.

Ashton, J., and H. Seymour. 1988. *The New Public Health.* Milton Keynes, UK: Open University Press.

Beck, U. 1992. *Risk Society.* London: Sage.

Bjerregaard, P., T.K. Young, E. Dewailly, and S.O.E. Ebbesson. 2004. Indigenous Health in the Arctic. *Scandinavian Journal of Public Health, 32,* 390-95.

Briggs, D., C. Corvalán, and M. Nurminen. 1996. *Linkage Methods for Environment and Health Analysis.* Geneva: WHO.

Brown, P. 1995. Race, Class and Environmental Health. *Environmental Research, 69,* 5-30.

–. 1992. Popular Epidemiology and Toxic Waste Contamination. *Journal of Health and Social Behavior, 33,* 267-81.

Brulle, R., and D. Pellow. 2006. Environmental Justice. *Annual Review of Public Health, 27,* 103-24.

Burger, E. 1990. Health as a Surrogate of Environment. *Daedalus, 11*(4), 133-53.

Burnett, R., S. Camak, and J.R. Brook. 1998. The Effect of the Urban Ambient Air Pollution Mix on Daily Mortality Rates in 11 Canadian Cities. *Canadian Journal of Public Health, 89,* 152-56.

Butler-Walker, J., J. Houseman, and L. Seddon. 2006. Maternal and Umbilical Cord Blood Levels of Mercury, Lead, Cadmium and Essential Trace Elements in Arctic Canada. *Environmental Research, 100,* 295-318.

Butler-Walker, J., L. Seddon, E. McMullen, J. Houseman, K. Tofflemire, A. Corriveau, J.P. Weber, C. Mills, S. Smith, and J. Van Oostdam. 2003. Organochlorine Levels in Maternal and Umbilical Cord Blood in Plasma in Arctic Canada. *Science of the Total Environment, 302,* 27-52.

Buzzelli, M. 2007. Bourdieu Does Environmental Justice? Probing the Linkages between Population Health and Air Pollution Epidemiology. *Health and Place, 13,* 3-13.

Buzzelli, M., and M. Jerrett. 2004. Racial Gradients in Ambient Air Pollution Exposure in Hamilton, Canada. *Environment and Planning A, 36,* 1855-76.

Canada. 2006. *Federal Budget Update.* Ottawa: Ministry of Finance.

–. 1986. *Achieving Health for All.* Ottawa: Government of Canada.

–. 1974. *A New Perspective on the Health of Canadians.* Ottawa: Government of Canada.

Centers for Disease Control and Prevention. 2005. *Social Determinants of Health.* Atlanta: Centers for Disease Control and Prevention.

Clark, G., S. Elliott, and J. Eyles. 2006. Lay Views of Disclosure in Post-Normal Times. In K. Andersson (Ed.), *Values in Decisions on Risk.* Stockholm: Swedish Nuclear Power Inspectorate.

Colburn, T., D. Dumanoski, and J. Peterson-Meyers. 1996. *Our Stolen Future.* New York: Dutton.

Cutter, S. (Ed.). 2001. *American Hazardscapes.* Washington, DC: Joseph Henry Press.

Delfino, R.J., A.M. Murphy-Moulton, R.T. Burnett, J.R. Brook, and M.R. Becklake. 1997. Effects of Air Pollution on Emergency Room Visits for Respiratory Illness in Montreal, Quebec. *American Journal of Respiratory and Critical Care Medicine, 155,* 568-76.

Dewailly, E., and P. Weike. 2003. The Effects of Arctic Pollution on Population Health in Arctic Monitoring and Assessment Program: *Human Health in the Arctic.* Oslo: Arctic Monitoring and Assessment Program.

Dunn, J. 2006. Speaking Theoretically about Population Health. *Journal of Epidemiology and Community Health, 60,* 572-73.

Elliott S., J. Eyles, and P. DeLuca. 2001. Mapping Health in the Great Lakes Areas of Concern. *Environmental Health Perspectives, 109*(suppl. 6), 817-26.

Environmental Defence. 2006. *Polluted Children, Toxic Nation.* Toronto: Environmental Defence.

–. 2005a. *Great Lakes, Great Pollution.* Toronto: Environmental Defence.

–. 2005b. *Toxic Nation.* Toronto: Environmental Defence.

Evans, G., and E. Kantrowitz. 2002. Socioeconomic Status and Health. *Annual Review of Public Health, 23,* 303-31.

Evans, R., M. Barer, and T. Marmor (Eds.). 1994. *Why Are Some People Healthy and Others Not?* New York: Aldine de Gruyter.

Evans, R., and G. Stoddart. 1990. Producing Health, Community Health Care. *Social Science and Medicine, 31,* 1347-63.

EWG (Environmental Working Group). 2006. *Across the Generations.* Washington: Environmental Working Group.

–. 2005. *Body-Burden: The Pollution in Newborns.* Washington, DC: Environmental Working Group.

Eyles, J. 2001. A Political Ecology of Environmental Contamination? In P. Armstrong, H. Armstrong, and D. Coburn (Eds.), *Unhealthy Times* (pp. 171-94). Don Mills: Oxford University Press.

–. 1999. Health, Environmental Assessments and Population Health. *Canadian Journal of Public Health, 90*(suppl. 1): S31-S34.

Eyles, J., S. Birch, and S. Chambers. 1994. Fair Shares for the Zone. *Canadian Geographer, 38,* 134-50.

Eyles, J., and N. Consitt. 2004. What's the Risk? *Environment, 46*(8), 24-39.

Finkelstein, M., M. Jerrett, P. DeLuca, N. Finkelstein, D.K. Verma, K. Chapman, and M.R. Sears. 2003. Relation between Income, Air Pollution and Mortality. *Canadian Medical Association Journal, 169,* 397-402.

Finkelstein M., M. Jerrett, and M.R. Sears. 2004. Traffic Air Pollution and Mortality Rate Advancement Periods. *American Journal of Epidemiology, 160,* 173-7.

Frank, J. 1995. Why Population Health? *Canadian Journal of Public Health, 86,* 162-5.

Frohlich, K., E. Corin, and L. Potvin. 2001. A Theoretical Proposal for the Relationship between Context and Disease. *Sociology of Health and Illness, 23,* 776-97.

Fuchs, M. 1996. Woburn's Burden of Proof. *Journal of Undergraduate Science, 3,* 165-70.

Furgal, C., and R. Keith. 1998. Canadian Arctic Contaminants Assessment Report. *Northern Perspectives, 25.* http://www.carc.org/pubs/v25no2/w252.htm.

Goldberg, M., R.T. Burnett, J. Brook, J.C. Bailar, M.F. Valois, and R. Vincent. 2001. Association between Daily Cause-Specific-Mortality and Concentration of Ground Level Ozone in Montreal, Quebec. *American Journal of Epidemiology, 154,* 817-26.

Haalboom B., S.J. Elliott, J. Eyles, and H. Muggah. 2006. The Risk Society at Work in Sydney Tar Ponds. *Canadian Geographer, 50,* 227-41.

Harr, J. 1996. *A Civil Action.* New York: Random House.

Health Canada. 2003. *The Social Determinants of Health: An Overview.* Ottawa: Health Canada.

–. 2001. *Social Determinants of Health.* Ottawa: Health Canada.

–. 2000. *A Statistical Profile of the Health of First Nations in Canada.* Ottawa: Health Canada.

–. 1999. *Addressing the Determinants of Health.* Ottawa: Health Canada.

Hertzman C., J. Frank, and R.G. Evans. 1994. Heterogeneities in Health Status and the Determinants of Population Health. In R. Evans, M. Barer, and T. Marmor (Eds.), *Why Are Some People Healthy and Others Not?* (pp. 67-92). New York: Aldine de Gruyter.

Hollander, D. 1997. Environmental Effects on Reproductive Health. *Family Planning Perspectives, 29,* 82-89.

Institute of Medicine. 2004. *Environmental Health Indicators.* Washington, DC: Institute of Medicine.

–. 1999. *Toward Environmental Justice.* Washington, DC: Institute of Medicine.

Irwin A., N. Valentine, C. Brown, R. Loewenson, O. Solar, H. Brown, T. Koller, and J. Vega. 2006. The Commission on Social Determinants of Health. *PLoS Medicine* (23 May). http://medicine.plosjournals.org/perlserv/?request=get-documentanddoi=10.1371%2F journal.pmed.0030106&ct=1.

Jerrett, M., R. Burnett, J. Brook, P. Kanaroglou, C. Giovis, N. Finkelstein, and B. Hutchison. 2004. Do Socioeconomic Characteristics Modify the Short Term Association between Air Pollution and Mortality? *Journal of Epidemiology and Community Health, 58,* 31-40.

Jerrett, M., M. Buzzelli, R.T. Burnett, and P.F. DeLuca. 2005. Particulate Air Pollution, Social Confounders and Mortality in Small Areas in an Industrial City. *Social Science and Medicine, 60,* 2845-63.

Jerrett, M., J. Eyles, C. Dufournaud, and S. Birch. 2003. Environmental Influences on Health Care Expenditures. *Journal of Epidemiology and Community Health, 57,* 334-38.
Jouannet, P., C. Wang, F. Eustache, T. Kold-Jensen, and J. Auger. 2001. Semen Quality and Male Reproductive Health. *APMIS, 109,* 333-44.
Katsouyanni, K. 2003. Ambient Air Pollution and Health. *British Medical Bulletin, 68,* 143-56.
Keller-Olaman, S., J.D. Eyles, S.J. Elliott, K. Wilson, N. Dostrovsky, and M. Jerrett. 2005. Individual and Neighbourhood Characteristics Associated with Environmental Exposure. *Environment and Behavior, 37,* 441-64.
Lindbladh, E., C.H. Lyttkens, B.S. Hanson, and P.O. Ostergren. 1998. Equity Is Out of Fashion? *Social Science and Medicine, 46,* 1017-25.
Litt, J., N.L. Tran, and T.A. Burke. 2002. Examining Urban Brownfields through the Public Health "Macroscope." *Environmental Health Perspectives, 110*(Suppl. 2), 183-93.
Mackenzie, C., A. Lockridge, and M. Keith. 2005. Declining Sex Ratios in a First Nation Community. *Environmental Health Perspectives, 113,* 1295-98.
McKeown, T. 1976. *The Role of Medicine.* Oxford: Nuffield Trust.
Murray, C., M. Ezzati, A.D. Lopez, A. Rodgers, and S. Vander Hoorn. 2003. Comparative Quantification of Health Risks. *Population Health Metrics, 1*(1). http://www.pophealthmetrics.com/content/1/1/1.
Northridge, M., G.N. Stover, J.E. Rosenthal, and D. Sherard. 2003. Environmental Equity and Health. *American Journal of Public Health, 93,* 209-14.
OCFP (Ontario College of Family Physicians). 2004. *Systematic Review of Pesticide Human Health Effects.* Toronto: Ontario College of Family Physicians.
OMA (Ontario Medical Association). 2005. *Illness Costs of Air Pollution.* (2nd ed.) Toronto: Ontario Medical Association.
–. 2000. *Illness Costs of Air Pollution.* Toronto: Ontario Medical Association.
O'Neill, M., M. Jerrett, I. Kawachi, J.I. Levy, A.J. Cohen, N. Gouveia, P. Wilkinson, T. Fletcher, L. Cifuentes, and J. Schwartz. 2003. Health, Wealth and Air Pollution. *Environmental Health Perspectives, 111,* 1861-70.
Ontario. 2006. *Ontario Budget 2006.* Toronto: Ministry of Finance.
Overpeck, J., K. Hughen, D. Hardy, R. Bradley, R. Case, M. Douglas, B. Finney, K. Gajewski, G. Jacoby, A. Jennings, S. Lamoureux, A. Lasca, G. MacDonald, J. Moore, M. Retelle, S. Smith, A. Wolfe, and G. Zielinski. 1997. Arctic Environmental Change of the Last Four Centuries. *Science, 278,* 1251-56.
Parry, J., and A. Stevens, 2001. Prospective Health Impact Assessment. *British Medical Journal, 323,* 1177-82.
Paulozzi, L. 1999. International Trends in Rates of Hypospadias and Cryptochiadism. *Environmental Health Perspectives, 107,* 297-302.
Petreas, M., J. She, F.R. Brown, J. Winkler, G. Windham, E. Rogers, G. Zhao, R. Bhatia, and M.J. Charles. 2003. High Body Burden of 2,2'-4,4'-Tetrabromodiphenyl (BDE-47) in California Women. *Environmental Health Perspectives, 111,* 1175-79.
PHAC (Public Health Agency of Canada). 2007. What Is the Population Health Approach? http://www.phac-aspc.gc.ca/ph-sp/phdd/approach/index.html.
Poikolanian, J., I. Tombre, and R. Kallenborn. 2004. Estimation of Long-Range Transport of Mercury, Cadmium and Lead to Northern Finland on the Basis of Moss Surveys. *Arctic, Antarctic and Alpine Research, 36,* 292-97.
Polar Research Board-Marine Research Board (PRB-MRB). 2006. *Polar Ice-Breakers in a Changing World.* Washington, DC: National Academies Press.
Raphael, D. 2004. *Social Determinants of Health.* Toronto: Scholars Press.
–. 2003. Addressing the Social Determinants of Health in Canada. *Policy Options,* March, 35-40.
Regidor, E. 2006. Social Determinants of Health. *Journal of Epidemiology and Community Health, 60,* 896-901.
Schecter, A., M. Pavuk, O. Päpke, J.J. Ryan, L. Birnbaum, and R. Rosen. 2003. Polybrominated Diphenyl Ethers (PBDEs) in US Mother's Milk. *Environmental Health Perspectives, 111,* 1175-79.

Schettler, T., G. Solomon, M. Valenti, and A. Huddle. 2000. *Generations at Risk.* Cambridge, MA: MIT Press.
Scott-Samuel, A. 1998. Health Impact Assessment. *Journal of Epidemiology and Community Health, 52,* 704-5.
Smith, K., and M. Ezzati. 2005. How Environmental Health Risks Change with Development. *Annual Review of Environment and Resources, 30,* 291-333.
Solomon, G., and T. Schettler. 2000. Environment and Health: 6 Endocrine Disruption and Potential Human Health Implications. *Canadian Medical Association Journal, 163,* 1471-76.
Solomon, G., and P. Weiss. 2002. Chemical Contaminants in Breast Milk. *Environmental Health Perspectives, 110,* A339-47.
Suk, W., M.D. Avakian, D. Carpenter, J.D. Groopman, M. Scammell, and C.P. Wild. 2004. Human Exposure Monitoring and Evaluation in the Arctic, *Environmental Health Perspectives, 112,* 113-20.
Suzuki Foundation. 1998. *Taking Our Breath Away.* Vancouver: Suzuki Foundation.
Taylor, D. 2000. The Rise of the Environmental Health Paradigm. *American Behavioral Sciences, 43,* 508-80.
Thompson, H., R. Atkinson, M. Petticrew, and A. Kearns. 2006. Do Urban Regeneration Programmes Improve Public Health and Reduce Health Inequalities? *Journal of Epidemiology and Community Health, 60,* 108-15.
Townsend, P., and N. Davidson. 1982. *Inequalities in Health.* London: Penguin.
Tsuji, L., B.C. Wainman, I.D. Martin, J.P. Weber, C. Sutherland, J.R. Elliott, and E. Nieboer. 2005. The Mid-Canada Radar Line and First Nations People of the James Bay Region, Canada. *Journal of Environmental Monitoring, 7,* 888-98.
Turner, B., R.E. Kasperson, P.A. Matson, J.J. McCarthy, R.W. Corell, L. Christensen, N. Eckley, J.X. Kasperson, A. Luers, M.L. Martello, C. Polsky, A. Pulsipher, and A. Schiller. 2003. A Framework for Vulnerability Analysis in Sustainability Science. *Proceedings of the National Academy of Sciences of the United States of America.* http://www.pnas.org/cgi/content/full/100/14/8074?maxtoshow=&HITS=10&hits=10&RESULTFORMAT=&fulltext=turner+vulnerability+analysis&searchid=1&FIRSTINDEX=0&resourcetype=HWCIT.
Villeneuve, P., R.T. Burnett, Y.L. Shi, D. Krewski, M.S. Goldberg, C. Hertzman, Y. Chen, and J. Brook. 2003. A Time-Series Study of Air Pollution, Socioeconomic Status and Mortality in Vancouver, Canada. *Journal of Exposure Analysis and Environmental Epidemiology, 13,* 427-35.
Vineis, P., and D. Kriebel. 2006. Causal Models in Epidemiology. *Environmental Health: A Global Access Science Source, 5*(21). http://www.ehjournal.net/content/5/1/21.
Wang, X., and J. Key. 2003. Recent Trends in Arctic Surface, Cloud and Radiation Properties from Space. *Science, 299,* 1934-37.
Wilkinson, R., and M. Marmot. 2003. *Social Determinants of Health: The Solid Facts.* (2nd ed.) Copenhagen: World Health Organization.
Williams, D., and C. Collins. 2001. Racial Residential Segregation. *Public Health Reports, 16,* 404-16.
Wolcott, C., K.J. Aronson, W.M. Hanna, S.K. Sengupta, D.R. McCready, E.E. Sterns, and A.B. Miller. 2001. Organochlorines and Breast Cancer Risk by Receptor Status, Tumor Site and Grade (Canada). *Cancer Causes and Control, 12,* 395-404.
WHO (World Health Organization). 2005. *Health Impact Assessment for Cities.* Copenhagen: World Health Organization
–. 2002. *The World Health Report.* Geneva: World Health Organization.
–. 1999. *Protection of the Human Environment.* Geneva: World Health Organization.
–. 1986. *Achieving Health for All.* Geneva: World Health Organization.
Younglai, E., J.A. Collins, and W.G. Foster. 1998. Canadian Semen Quality. *Fertility and Sterility, 70,* 76-80.
Zanobetti, A., and J. Schwartz. 2000. Race, Gender and Social Status as Modifiers of the Effects of PM_{10} on Mortality. *Journal of Occupational and Environmental Medicine, 42,* 409-74.

8
Environmental Injustice in the Canadian Far North: Persistent Organic Pollutants and Arctic Climate Impacts

Sarah Fleisher Trainor, Anna Godduhn, Lawrence K. Duffy, F. Stuart Chapin III, David C. Natcher, Gary Kofinas, and Henry P. Huntington

Although residents of the Canadian Far North are at the geographic, demographic, and economic periphery of Canada, they stand in the centre of two major forms of environmental injustice: exposure to toxic persistent organic pollutants (POPs) and impacts of a changing Arctic climate. These issues share a common feature with each other and with many instances of environmental injustice: while negative impacts are experienced locally, the industrial source is distant – in this case, thousands of kilometres away in southern latitudes including sources outside Canada – making them inherently cross-scale and international issues (Williams, 1999).[1] During international treaty negotiations in the late 1990s and early 2000s, the Inuit Circumpolar Conference (ICC) and other northern indigenous groups prominently framed their incidental exposure to POPs in both public health and human rights terms and were largely successful in achieving their goals of international regulation of POPs. More recently, the ICC has been among the first to frame Arctic climate impacts as an issue of human rights.

This chapter highlights the issues of toxic POPs exposure and Arctic climate impacts as issues of environmental injustice in Canada, noting the following key findings. First, political and industrial activity on national and global scales can have momentous environmental, social, and cultural repercussions on the local scale in remote areas. Second, solutions to these cross-scale problems require local adaptation as well as strategic political activity in national and international arenas. We further discuss the ways in which strategies of increasing adaptive capacity and decreasing vulnerability may complement mitigation and allow for comprehensive community sustainability that includes social, cultural, economic, and environmental components.

Less than 1 percent of Canada's total population lives in the Canadian

North (Nunavut, Northwest, and Yukon territories). Roughly 4 percent of Canadian gross domestic product (GDP) occurs in this region, the bulk of which originates in the Northwest Territories; Nunavut alone produces less than 1 percent of Canadian GDP.[2] Nation-wide Aboriginal people (Inuit, First Nations, and Métis) constitute only 3 percent of the Canadian population, yet they represent over half of the population in the Far North.[3] They live in communities ranging from one hundred to several thousand people and rely on marine mammals, fish, birds, caribou, moose, small game such as ptarmigan and snowshoe hare, and various types of berries for a large percentage of their nutritional requirements. Aside from nutritional and economic value, the procurement of wild foods is deeply tied to cultural identity (Berkes, 1999; Duhaime, 2002; Furgal and Seguin, 2006).

We consider the problem of environmental injustice, conceived more broadly as environmental inequality, to be one in which some people bear disproportionate environmental burdens of industrial by-products or otherwise have inequitable access to environmental goods and services (Pellow, 2000).

There is abundant evidence, acquired both via Western science and traditional knowledge, that indigenous peoples and ecosystems of northern Canada and throughout the Arctic are experiencing significant health impacts from the accumulation of POPs and rapid and notable ecological, social, and cultural impacts from a changing climate (Krupnik and Jolly, 2002; AMAP, 2003; Downie and Fenge, 2003; Berner and Furgal, 2005; Chapin et al., 2005a; Huntington and Fox, 2005; McCarthy and Martello, 2005). The Inuvialuit, Gwich'in, Dene, Inuit, Innu, and other groups rely on an intimate relationship with their local ecosystem for physical and cultural sustenance. As culture groups, they are each dynamic, learning entities and have for millennia adapted to many changes. The extent to which northern indigenous communities and cultures will be able to adapt to the combined effects of a polluted food web and the rapid changes related to climate change remains unclear (Kinloch, Kuhnlein, and Muir, 1992; Duerden, 2004; Myers, Fast, Berkes, and Berkes, 2005; Myers and Furgal, 2006; Cone, 2005). Adding to this uncertainty are the impacts of contaminants and climate change on indigenous resource harvesters of the future.

The case of Arctic climate impacts raises an additional important issue of *intra*-national equity with respect to climate change, which has received little scholarly attention. The vast majority of literature on issues of climate change and equity and justice engage *inter*national negotiations and revolve around the question of equitable global distribution of greenhouse gas emissions as related to economic development in the Third World (see, for example, Adger, 2001; Jamieson, 2001; Agarwal, Narain, and Sharma, 2002; Pinguelli-Rosa and Munashinghe, 2002).[4] In contrast, the issue of climate impacts in the Canadian and American Arctic and Subarctic raises the question of unequal distribution of burdens and benefits *within* more developed, or First World, countries. These impacts can be characterized as aspects of internal colonialization – the colonization and exploitation of those living in the hinterlands of developed countries (Osherenko and Young, 1989).

Persistent Organic Pollutants

Organic chemical synthesis of items such as medicines, pesticides, lubricants, solvents, adhesives, fabrics, textiles, and plastics has grown and continues to expand, adding convenience and safety to people's lives (Thornton, 2000). However, waste streams from the manufacture and consumption of thousands of new organic and biologically active substances tend to carry persistent and toxic compounds (Thornton, 2000; Schultz, 2004). Bonds between halogen elements and carbon thwart metabolism and have environmental half lives of months, years, or decades. These compounds, known as persistent organic pollutants (POPs), are fat soluble and bioaccumulate in organisms, with upper food-chain carnivores such as polar bears and humans having highest concentrations (Suedel, Boraczek, Peddicord, Clifford, and Dillon, 1994; Borga, Gabrielsen, and Skaare, 2001).

Although military and communications activities have generated point-source contamination throughout northern Canada and the Arctic (Reiersen, 2000; McCreanor, 2003), long-range transport and bio-magnification in the food chain is a significant source of POPs contamination in humans in the Arctic (AMAP 2003, 2004). Long-range transport occurs mostly in the atmosphere, via dust particles that travel trans-globally in a matter of days in a process known as the grasshopper effect. Rising with the warm polluted air of temperate and tropical regions and settling as air moves north and cools, persistent, oily pollutants such as DDT (the pesticide dichloro-diphenyl-trichloroethane), PCBs (polychlorinated biphenyls), and PCDDs (polychlorinated dibenzo dioxins) are produced and utilized in southern latitudes, yet migrate to and accumulate in the Arctic (Godduhn and Duffy 2003). To lesser and less-well-studied degrees, ocean currents and migratory species also carry pollution toward the poles (AMAP, 1998). The presence of human-made chemicals throughout the Arctic necessarily engages the issue of environmental justice on a global scale, illustrating the tragedy of "the [global] commons" (Middaugh, 2001, p. 242).

Although POPs are a contaminant and health risk throughout the world, indigenous people in the Arctic are exposed to a disproportionate burden of non-point source POPs contaminants because of their location at the highest trophic level of the Arctic food web. The breast milk fat of Inuit women in the Canadian Arctic have two to ten times greater organochlorine concentrations than samples from Caucasian women in southern Quebec and higher mean polychlorinated biphenyl (PCB) concentration levels than those reported for general female populations worldwide. In fact, the greatest body burden known to occur from environmental exposure is found in Inuit mothers (Dewailly et al., 1993, p. 620).[5]

Arctic-wide blood concentration levels of PCBs in the Inuit from Baffin and Nunavik in eastern Canada were third highest, surpassed only by those in the west and east coasts of Greenland. PCB levels exceeding what Health Canada considers to be a level of concern (above 5 µg/L) were found in 43 percent of the samples from Inuit women from the Northwest Territories (NWT) and Nunavut. In Baffin, 73 percent of the population exceeded this national standard for PCBs in blood and nearly 60 percent of the populations of Kivalliq and Nunavik exceeded the standard. In these locations, Dene/Métis and Caucasian values were 3.2 percent and 0.7 percent respectively. The population in Kitikneot was recorded with over 40 percent exceedence, and that of Inuuvik at roughly 15 percent exceedence (AMAP, 2003, p. 103).[6] In addition, while not necessarily directly linked with POPs, compared to statistics for general populations in North America and Greenland, indigenous peoples experience consistently lower life expectancy at birth, higher rates of infectious disease, and higher rates of mortality from accidents and suicide (AMAP, 2003; Van Oostdam et al., 2005)

The known health effects of POPs exposure include diverse, sublethal effects such as life-long alterations to immune, reproductive, and neurological function. Research in the Arctic has helped confirm that exposure to POPs in the womb and early childhood is a risk factor in rates of infections, learning disabilities, and the formation of abnormal sex organs, with additional potential links to cancer and osteoporosis (Dewailly et al. 1993; Andersson, Girgor, Meyts, Leffers, and Skakkebæk, 2001). In addition, psychological impacts of POPs contamination in the Arctic have profound impacts on food security and the longevity of cultural traditions involving hunting, fishing, and food preparation (Hild, 1998, 2003; Van Oostdam et al., 1999; Watt-Cloutier, 2003). The presence of known carcinogens and other harmful toxins in the traditional country food supply generates a sentiment of fear. Although Arctic indigenous people place a high value on maintaining cultural traditions and teaching to youth traditional knowledge related to hunting, fishing, and food preparation, they do not wish to knowingly expose themselves and their families to harmful chemicals (Tyrrell, 2006). Furthermore, indigenous people in the Arctic have a high degree of respect for the plants, animals, and landscape in which they live and upon which they depend for survival. They view plants and animals as moral equivalents to humans and see themselves as an intimate part of this interconnected system (Hild, 2003). This psychological burden is described by Sheila Watt-Cloutier, then president of Inuit Circumpolar Council Canada, in her testimony at POPs treaty negotiations: "Imagine for a moment if you will the emotions we now feel – shock, panic, rage, grief, despair – as we discover that the food which for generations has nourished us and keeps us whole physically and spiritually is now poisoning us. You go to the supermarket for food. We go out on the land to hunt, fish, trap and gather. The environment is our supermarket . . . As we put our babies to our breasts we feed them a noxious chemical cocktail that foreshadows neurological disorders, cancer, kidney failure, reproductive dysfunction, etc. This is truly worrying."[7]

Arctic Climate Impacts

There is strong scientific consensus that the rapid increase in fossil fuel combustion associated with the industrial revolution has contributed significantly to the direction and strength of the climate signal (Crowley, 2000; National Research Council, 2001). Although the bulk of fossil fuel combustion occurs in temperate and tropical regions, the warming is amplified at high latitudes, where changes in vegetation and reductions in sea ice, snow cover, and glaciers have increased absorption of solar energy and atmospheric heating (Anisimov et al., 2001; Chapin et al., 2005a; McBean et al., 2005). Thus, when considering impacts on the immediate time horizon (i.e., the next two to twenty years), fossil fuel combustion and international climate policy on a global scale currently impact northern people disproportionately. Although some climate impacts may be advantageous, indigenous peoples in the North are likely to gain less and lose more than people residing to the south and are therefore the relative, if not absolute, "losers" in terms of current impacts of climate change (O'Brien and Leichenko, 2003).

The earth's climate is changing. The biophysical manifestations of these changes are most prominent in high latitudes, and the rate of Arctic change is projected to increase over time. In the past few decades, the average Arctic temperature has risen at nearly twice the rate of the rest of the world. On the ground, this has been observed as warmer spring and autumn temperatures that are associated with changes in seasonality, including earlier ice thaw and later freeze-up. In some cases, these phenomena pose significant hazards for transportation in hunting, as well as diminishing habitat for marine mammals and polar bears. Decreased shorefast ice and sea-ice extent have led to increased storm intensity, causing severe coastal erosion and requiring the relocation of homes and communities. Lakes are drying up and water temperatures are rising in lakes and rivers, affecting fish populations. Weather patterns that have for generations served as cues for hunters in accessing game and travelling safely are no longer familiar or predictable. The physical, social, and cultural impacts from climate change are projected to intensify over time as the climate continues to warm (Serreze et al., 2000; Krupnik and Jolly, 2002; Lynch, Curry, Brunner, and Maslanik, 2004; Chapin et al., 2005b; Hinzman et al., 2005).

Climate change is impacting the cultural and economic activities of indigenous peoples in the Arctic, which are closely tied to the physical and biological aspects of the Arctic environment. Arctic people have survived for generations in close relationship with their environment through keen observation, flexibility, and adaptation. Forces internal to northern communities also generate social and cultural change, and not all change is detrimental nor unwanted (e.g., arrival of new species might provide new food sources). However, the relatively rapid rate of biophysical change associated with a changing climate challenges people's capacity to respond and adapt (Kofinas et al., 2005; Leiserowitz, Gregory, and Failing, 2006). The resilience of northern indigenous communities and of the social ecological system of the Arctic is thus strongly influenced by a changing climate (Berkes and Jolly, 2001; Chapin et al., 2004).

For example, the Inuit of Nunavut rely on the ringed seal as a staple food source year-round. Ringed seals and polar bears depend on stable shorefast sea ice for breeding. Over the past thirty years, the average extent of sea ice has declined by 15 to 20 percent (Serreze et al., 2000). Modest climate change scenarios project an acceleration of this melting with a nearly ice-free Arctic Ocean within the next one hundred years (Overpeck et al., 2005). In May 2008, the United States Fish and Wildlife Service listed the polar bear as a threatened species because of this loss of habitat (Weiss, 2008). As the Arctic Climate Impact Assessment (2004) states, "To hunt, catch, and share these foods is the essence of Inuit culture. Thus the decline in ringed seals and polar bears threatens not only the dietary requirements of the Inuit, but also their very way of life . . . Because ringed seals and polar bears are very unlikely to survive in the absence of summer sea ice, the impacts on indigenous communities that depend on these species is likely to be enormous" (p. 94).

Peter Irniq, commissioner of Nunavut, describes climate change impacts on the Inuit in his territory:

> Our cultural, physical and economic survival depended on the availability of abundant and healthy stocks of wildlife and fishes. We were always aware of the need to protect our environment and aware of what consequences there could be if we did not do just that. Since time immemorial we have harvested and at the same time managed wildlife according to our own laws of wildlife and conservation based on respect of nature . . . Our whole structure depended on the passing of the seasons, the predictability of the weather, the direction of the wind, and the thickness of the ice, the *nuuttuittuq*, the North Star . . . We adapted to changes that occurred over

long periods of time, like natural migration changes. What we are facing today are drastic changes, happening within decades . . . There is a real threat to our way of life because of climate change and global warming. As Inuit, we are already experiencing both positive and negative impacts of climate change. (2004, n.p.)

The Inuvialuit of the western Canadian Arctic initiated a study of climate change impacts in the village of Sachs Harbour to document the significant environmental changes that are occurring and their social and cultural impacts. These include the following impacts:

- lakes draining into the sea from permafrost thawing and ground slumping
- less sea ice in summer and accompanying rougher water
- weather, storms, and spring breakup difficult to predict
- winter travel dangerous because of unpredictable sea ice and excessive broken ice
- less ice cover in summer, leading to rougher, more dangerous storms at sea
- unstable ice, making hunting more difficult in winter
- harder to hunt geese because the spring melt happens so fast
- presence of fish and bird species that have never been seen before
- increase of biting flies and mosquitoes. (ACIA, 2004, p. 95)

Climate change has similar impacts on Iñupiat in northern Alaska, Yupik in western Alaska (Krupnik and Jolly, 2002), and other indigenous peoples of the Arctic. In Canada and throughout the Arctic, indigenous people are documenting the biophysical, nutritional, cultural, and social impacts of climate change. These observations are consistent with the findings of Western science-based climate change research (Hinzman et al., 2005)

Issue Framing and Political Empowerment

Arctic climate impacts and POPs accumulation are thus two examples of environmental injustice in northern Canada. Both cases involve people in remote regions who depend on wild foods and bear a physical and cultural burden that is disproportionate to the benefits they receive from industrial production and consumption. The evidence for these conditions is supported by rigorous and comprehensive scientific studies sponsored by the Arctic Council, an international body of national representatives from all Arctic nations. The work directed by the Arctic Council has provided scientific foundation for policy debate, raised awareness of pertinent local issues, and influenced problem framing (AMAP, 2003; ACIA, 2004; Young,

2005). In both instances, local and Arctic-wide health and well-being are impacted by industrial activity in lower latitudes, raising the need for strategic cross-scale interactions and international activity (Kurtz, 2003; Berkes Bankes, Marschke, Armitage, and Clark, 2005a). In this section we consider the problem frames of these issues and the role of indigenous people in shaping political action.

Strategies and Frames: POPs

Throughout the period of treaty negotiations and political action, POPs contamination in the Canadian Far North was framed primarily as an issue of public health (ICC, 2005). Sheila Watt-Cloutier further framed the POPs issue as one of human rights. This underscores the perspective of Aboriginal peoples in the Arctic that environmental health is synonymous with public health, which in turn is intimately linked to the health and vibrancy of cultural traditions (Watt-Cloutier, 2003). Although by assigning the Department of Foreign Affairs and International Trade as delegates in international POPs negotiations the Canadian government tried to frame the issue in economic and sustainable development terms, the public health component of the issue is most compelling to indigenous Arctic peoples and ultimately received the most attention and generated the most leverage in treaty negotiations. In alliance with the Canadian Inuit, the Canadian government readily provided $20 million "to assist in convention implementation" (Fenge, 2003, p. 205) and led the way by ratifying the Stockholm Agreement on the day it was signed.

The Northern Contaminants Program (NCP), funded by the federal government as part of the 1991 Arctic Environmental Strategy, was instrumental in "making the scientific case for international action and in persuading the government of Canada to argue for international agreements to control, reduce and eventually eliminate key POPs" (Fenge, 2003, p. 210). Ultimately, it played a key role in funding northern indigenous peoples, "allowing them to act effectively on the circumpolar and global stages, defending their interests and promoting their ways of life" (Fenge, 2003, pp. 210-11).

Coalition and collaboration are an important part of the success of the POPs treaty negotiations (Buccini, 2003; Fenge, 2003; Watt-Cloutier, 2003). During treaty negotiations, northern indigenous people worked together in coalition and "nurtured working links" with nongovernmental organizations, industry, and the Canadian delegation (Watt-Cloutier, 2003, p. 263). Organized initially as the Northern Aboriginal Peoples Coordinating Committee on POPs, this coalition later became known as the Canadian Arctic Indigenous Peoples Against POPs. Described by Fenge (2003) as precedent setting, this potent collaborative coalition of indigenous people from northern Canada, including Inuit, First Nation (Gwich'in/Dene) and Métis with strong leadership, was able to "exert" "real and acknowledged influence" on international treaty negotiations (p. 211).

Furthermore, the Inuit had a larger vision of collective consideration. Rather than taking a not-in-my-backyard approach in treaty negotiations, they presented POPs contamination as a global problem. As Sheila Watt-Cloutier explains, "If we can help people to see that a poisoned Inuk child, a poisoned Arctic and a poisoned planet are one in [sic] the same, then we will have effected a shift in people's awareness that will result without doubt in positive change."[8] This collective consideration is also illustrated in the Inuit stand toward the use of DDT to protect against malaria in lower latitude Third World countries. The issue of equity in access to public health was prominent in the POPs international negotiations, and the Inuit Circumpolar Conference (ICC) fought for public health for all peoples, advocating that alternative methods of malaria control be found. The Inuit framed POPs contamination in terms of their physical health and cultural survival while at the same time standing in solidarity for the health and well-being of people in other parts of the world and for the health of the ecosystem as a whole (Fenge, 2003). In this way, they view themselves as an important and significant piece of a larger global system that includes all people in intimate connection with the land, plants, animals, and environment (Hild, 2003).

Strategies and Frames: Arctic Climate Impacts

As with POPs contamination, the ICC is raising the political profile of regional and local Arctic climate impacts, framing the issue explicitly as one of human rights and climate justice (Watt-Cloutier, Fenge, and Crowley, *n.d.*; ICC, 2005). In December 2005, the ICC filed a petition with the Inter-American Commission on Human Rights, requesting assistance "in obtaining relief from human rights violations resulting from the impacts of global warming and climate change caused by acts and omissions of the United States" (ICC, 2005, p. 1). In this strategic appeal to an international body, the ICC takes a national government to task for its significant contribution to a global phenomenon, the impacts of which are felt on a local and regional level.

Terry Fenge, strategic counsel to the ICC during POPs negotiations, finds striking parallels between POPs contamination and climate change in the Arctic, both as "issue[s] of cultural survival to the Inuit" (2003, p. 211) as well for the Athabascan and other northern indigenous peoples. Fenge suggests that the Arctic Council, in conjunction with indigenous peoples, should "draw upon the POPS experience to re-balance popular perceptions by giving global climate change a human face" (p. 211). It is thus reasonable to infer that the recent framing of Arctic climate impacts by the ICC as a matter of climate justice is a strategic framing, built upon the successful negotiation of the 2001 United Nation's Stockholm Convention on POPs.

Sustainability, Adaptation, and Environmental Justice

There is growing recognition in the scientific community of the links between poverty, equity, and environmental sustainability and the need to address these issues simultaneously in policy solutions (Dresner, 2002; Sachs and Reid, 2006). As instances of environmental injustice in northern Canada, POPs contamination and impacts from climate change are similar in that they both have globally dispersed causes, decadal longevity in the environment, and large cultural impacts on northern peoples. As noted above, these cases exemplify the importance of cross-scale interactions and strategic political action in striving for environmental justice (Kurtz, 2003; Berkes et al., 2005a). Because social, economic, political, and environmental conditions vary throughout the North, one of the most promising approaches for addressing sustainability needs of indigenous peoples is through monitoring and assessing vulnerabilities, adaptive capacity, and resilience of specific locales (Berkes and Jolly, 2001; Smit and Pilifosova, 2001; Berner and Furgal, 2005; Huntington and Fox, 2005; McCarthy and Martello, 2005; Chapin et al., 2006).

In highlighting the inherent connections between sustainability and social equity and calling for a "broad-based political movement for ecological democracy," Faber and McCarthy (2003, p. 61) distinguish between social action that aims toward distributional justice and social action that aims toward productive justice. Distributional justice hinges on the demographic and geographic distribution of harmful industrial by-products and beneficial environmental goods. It is this form of justice for which the not-in-my-backyard (NIMBY) struggles strive. In contrast, productive environmental justice focuses on the initial production of harm. These struggles challenge the "systemic causes of social and ecological injustices as they exist in 'everyone's backyard'" (p. 60). Because in the cases of long-range POPs transport and climate impacts the source of contamination lies thousands of kilometres away from the afflicted peoples, the question of distributional justice assumes international and global dimensions. This necessitates a vision of collective consideration in which "my backyard" is essentially "everyone's backyard" and the issue of productive injustice overshadows that of distributional injustice.

As discussed above, indigenous peoples in northern Canada maintain mixed cash-subsistence economies. With high rates of poverty, few opportunities for wage earning, and high transportation costs associated with lack of road and rail access, they derive wealth from the natural abundance of plants, animals, and other resources upon which they directly depend for nutrition, and from their cultural traditions, which in turn are also intimately linked to the ecological system. For these Canadians, sustainability involves continuing subsistence harvest (Nuttall, 2000; Thompson, 2005, p. 62), and cultural and physical health is synonymous with environmental and ecosystem health. The longevity and integrity of cultural traditions are thus intimately linked to the integrity and health of a functioning ecosystem. In this way, productive justice for the Aboriginal peoples in northern Canada and throughout the Arctic is synonymous with ecological sustainability, which in turn is synonymous with cultural and social sustainability (Agyeman, Bullard, and Evans, 2003).

POPs contamination and climate change in the Arctic may be different from other, more localized cases of environmental injustice because mitigation (i.e., productive justice) alone cannot solve these problems. Each requires mitigation in conjunction with adaptation. Enhancing social ecological resilience and transformability are also important strategies for rectifying these environmental injustices (Chapin et al., 2006). Although mitigation is an essential component of remedying productive injustice, multidecadal residence times of food-web contaminants and atmospheric greenhouse gases makes this solution slow and uncertain, while threats to human and ecosystem well-being are large and immediate. Conversely, although both short- and long-term localized adaptation strategies may be required to avoid physical human harm, repairing these environmental injustices also requires mitigation of the offensive pollutants (Berkes and Jolly, 2001). Mitigation and adaptation should be considered two synergistic prongs of an attack on issues of productive injustice and not as alternative solutions to the problem.

In the case of POPs contamination, although the mitigation that resulted from the Stockholm Convention has been effective in reducing the use of toxins, their longevity in the environment and the continual development and production of new toxic chemicals cause POPs exposures to remain a health risk in northern Canada. Because adaptation to toxins in their food supply may require eliminating the consumption of certain wild foods and increasing consumption of store-bought foods, which would compromise both nutrition and cultural traditions, scientists and health professionals have generally recommended that the Inuit and other northern indigenous peoples continue to harvest and consume country foods on the grounds that, while contaminated, traditional foods are more nutritious than store-bought foods and the cultural and social benefit of maintaining traditional hunting, fishing, and gathering practice outweighs the biophysical harm (Northern Contaminants Programme, 1997; Van Oostdam et al., 2005).

However, more nuanced strategies for reducing toxic exposure may exist. Adaptation involves learning, communicating, experimenting, and innovating (Chapin et al., 2004). The Arctic Monitoring and Assessment Programme

(AMAP) has made huge strides in learning about contaminants, and about the geographic, ecological (i.e., species variations among food sources), age, and gender distributions of exposure and sensitivity. More detailed research is needed, however, to identify particularly vulnerable groups and detail differences in contaminant exposure. For example, up to a twofold difference in body burden of PCBs has been documented based on sex and age of marine mammals such as ringed seals and polar bears (Bard, 1999).

While the holistic world view of indigenous peoples may conflict with recommendations to avoid specific organs or consume only birds or animals of a certain age or sex, adaptation strategies developed in close partnership with the community can reduce toxic exposure while maintaining wild food consumption (Hild, 2003). For example, hunters may shift to harvest of young, male ring seals because their body burden of toxins is lower than that of older females. Similarly, women may avoid eating marine mammals when they are pregnant and nursing, infants may avoid consumption during their most rapid growth phase, and hunters can avoid sites that are affected by point-source contamination. Another aspect of adaptation is communication, networking, and developing polycentric governance structures to share information about contaminants and contaminants options. Given that clear communication of risk from scientist to community members is often lacking (Myers et al., 2005; Myers and Furgal, 2006), each of these potential strategies requires specific localized scientific research conducted in close partnership with communities. Most important in both POPs and climate impacts is that this process be one that empowers local people to learn and make decisions (Berkes, Huebert, Fast, Manseau, and Diduck, 2005b).

Similarly, rectifying the injustice of Arctic climate impacts will require a combined strategy of mitigation and adaptation. The biophysical implications of climate change in northern latitudes are self-reinforcing and, like POPs, many of them are irreversible on a time scale of from multiple decades to centuries. Measures should be taken to reduce greenhouse gas emissions worldwide, and people living in close association with their environment in the North must adapt to changes in their environment. The key to rectifying the injustice is to empower and engage the knowledge, priorities, and vision of local people on multiple scales of decision-making (Myers et al., 2005; Furgal and Seguin, 2006). For example, this can be achieved at the local scale through co-management institutions, at the territorial scale through land claims agreements and joint policy making, and at the international scale through international organizations and collaborations such as the ICC involvement in the Stockholm Convention on POPs (Nuttall, 2000, 2001; Berkes et al., 2005a).

Furthermore, additional work is needed to understand the synergistic impacts of multiple environmental, social, economic, and political stressors on northern indigenous peoples. With Arctic Council support, the Arctic Monitoring and Assessment Programme and the Arctic Climate Impact Assessment provide scientific documentation for POPs exposure and Arctic climate impacts. ICC is actively engaged in the political process to push for reduced greenhouse gas emissions in the United States. The Canadian government is starting to consider the need to provide resources for enhancing the resilience and adaptive capacity of northern indigenous peoples in the face of climate change (Peters, 2002; Furgal and Seguin, 2006). Relatively little is known, however, about the cumulative or synergistic impacts of contaminants and climate change or about interactions and feedbacks with social, economic, and political change (Duerden, 2004; Ebi, Kovats, and Menne, 2006).

In addition, we suggest that the disproportionate biophysical impacts from POPs contaminations and climate change be viewed in the larger context of social and cultural sustainability in northern indigenous communities and that people in these communities be directly involved in scientific monitoring and assessment as well as in policy decisions that chart a course of sustainability (Burger et al., 2006). Whereas the reductionist perspective and methods of Western science and public policy often lead to specific and issue-focused problem solving, northern indigenous people experience and understand causal links, their environment, and environmental change as interconnected with social, economic, political, and cultural conditions and change (Morrow, 1996; Huntington et al., 2006; Tyrrell, 2006). The ability of northern peoples to adapt to contamination and climate change is constrained by multiple stressors from economic, cultural, political, and environmental spheres, including pressures of colonization, assimilation, and rapid modernization. Connecting to the "wisdom of the land" and obtaining their food from the land are important ways they regain confidence, strength, and choices of life "over self destruction" and suicide (Watt-Cloutier, 2003, p. 257; Jilek, 1987). Solutions to the problems of POPs and climate change in the Arctic will best serve the communities if we begin to understand the links and potential feedbacks between these stressors and conceive of innovative solutions that directly involve community participation and address the community as a whole (Dewailly and Furgal, 2003; Furgal, 2005; Kraemer, Berner, and Furgal, 2005; McCarthy and Martello, 2005; Van Oostdam et al., 2005; Furgal and Seguin, 2006).

Conclusion

Toxic contamination from POPs and Arctic climate impacts are two examples of environmental injustice in northern Canada that involve local burdens deriving from national- or global-scale activities with longevity in the environment and large cultural impacts on northern peoples. Solutions to these problems thus require cross-scale interactions, and strategic political action across multiple scales. Both cases are scientifically documented in comprehensive studies (AMAP, 2003; ACIA, 2004) that were sponsored by the Arctic Council and, like many cases of environmental injustice, both illustrate the intricate connections among humans, the natural system, and the political process. There are indications that the Inuit Circumpolar Conference has drawn from its success with international negotiations of POPs in implementing ongoing strategies to frame Arctic climate impacts as an issue of human rights (Fenge, 2003).

The struggles of indigenous people in northern Canada against the environmental injustice of climate change and POPs contamination offer an exemplar for the integration of sustainability and justice. Solutions to these cases of injustice require both mitigation at the source and local adaptation involving strong local engagement in decision making at multiple scales. This will necessarily involve gaining greater understanding of the feedbacks and synergies among multiple stressors, including biophysical change; social, economic, and political impacts of assimilation; and rapid modernization. By considering social, cultural, economic, and ecological sustainability as inherently and intimately connected, efforts to rectify the environmental injustice of pollutants and climate impacts in northern Canada have the potential to take the lead in crafting innovative solutions that simultaneously cross scales, from local to global, and address the communities and ecosystems as a whole.

Notes

Portions of this chapter are excerpted from S.F. Trainor, F.S Chapin III, H.P. Huntington, G. Kofinas, D.C. Natcher (2007) Arctic Climate Impacts and Cross-Scale Linkages: Environmental Justice in Canada and the United States, *Local Environments: International Journal of Justice and Sustainability, 12*(6). We are grateful to Taylor and Francis Journals for its permission to partially reproduce this article and to La'Ona DeWilde for her insights on linking programs that enhance the ecological, social, and cultural resilience of northern Native communities. Funding for this project came from NSF Office of Polar Programs #0328282.

1 Notwithstanding the military sites and the emergence of a northern industrial economy (e.g., diamonds, oil and gas, and other minerals).
2 Statistics Canada, http://www40.statcan.ca/l01/cst01/econ15.htm?sdi=gross%20domestic%20product.
3 Statistics Canada, 2001 Census, http://www12.statcan.ca/english/census01/products/highlight/Aboriginal/Page.cfm?Lang=E&Geo=PR&View=1a&Table=1&StartRec=1&Sort=2&B1=Counts01&B2=Total.
4 Exceptions include S. Nishioka (1999), Fairness and Local Environmental Concerns in Climate Policy, in F. Toth (Ed.), *Fair Weather? Equity Concerns in Climate Change,* pp. 133-44 (London: Earthscan Publications); D.L. Peterson and D.R. Johnson (Eds.) (1995), *Human Ecology and Climate Change, People and Resources in the Far North* (Washington, DC: Taylor and Francis); Watt-Cloutier et al., n.d; and the Environmental Justice and Climate Change Initiative, http://www.ejcc.org/.
5 Additional long-range pollutants in the Arctic include heavy metals, such as methyl

mercury and acidifying compounds. Radionuclide contamination has resulted from external sources, such as atmospheric nuclear weapons testing (1950-1980), releases from nuclear reprocessing plants in England, and fallout from the Chernobyl accident in 1986, as well as from localized military and civilian sources. Radionuclides also bioaccumulate in the food chain, with consumers of wild, traditional foods having seven times greater doses than people who do not consume traditional foods (Reiersen, 2000, p. 577; Van Oostdam et al., 1999). See also L. Barrie, D. Gregor, B. Hargrave, R. Lake, D. Muir, R. Shearer, B. Tracey, and T. Bidleman (1992), Arctic Contaminants: Sources, Occurrence and Pathways, *Science of the Total Environment, 122,* 1-74.

6 While not classified as a POP, a similarly high concentration of mercury was found among the eastern Canadian Inuit, second only to Inuit in Greenland. By Health Canada guidelines (20 µg/L), 15 percent of the Nunavik population exceed acceptable mercury blood concentrations. This number jumps to near 80 percent if the 1999 standards set by the US Environmental Protection Agency are used (5.8 µg/L) (AMAP, 2003; Van Oostdam et al., 1999).

7 S. Watt-Cloutier (1999), The Need for a Global Treaty on Persistent Organic Pollutants, *Silarjualiriniq,* October-December, as cited in Fenge (2003, pp. 202-3).

8 Ibid.

References

ACIA (Arctic Climate Impact Assessment). 2004. *Impacts of a Warming Climate: Arctic Climate Impact Assessment.* Cambridge, UK: Cambridge University Press.

Adger, W.N. 2001. Scales of Governance and Environmental Justice for Adaptation and Mitigation of Climate Change. *Journal of International Development, 13,* 921-31.

Agarwal, A., S. Narain, and A. Sharma. 2002. The Global Commons and Environmental Justice: Climate Change. In J. Byrne, L. Glover, and C. Martinez (Eds.), *Environmental Justice* (pp. 171-99). New Brunswick, NJ: Transaction Publishers.

Agyeman, J., R.D. Bullard, and R. Evans (Eds.). 2003. *Just Sustainabilities, Development in an Unequal World.* Cambridge, MA: MIT Press.

AMAP (Arctic Monitoring and Assessment Programme) (Ed.). 2004. *AMAP Assessment 2002: Persistent Organic Pollutants in the Arctic.* Oslo: Arctic Monitoring and Assessment Programme.

–. 2003. *Human Health in the Arctic.* Oslo: Arctic Monitoring and Assessment Programme.

–. 1998. *Assessment Report: Arctic Pollution Issues.* Oslo: Arctic Monitoring and Assessment Programme.

Andersson, A.-M., K.M. Girgor, E.R.-D. Meyts, H. Leffers, and N.E. Skakkebæk (Eds.). 2001. *Hormones and Endocrine Disrupters in Food and Water: Possible Impact on Human Health.* Copenhagen: Munksgaard.

Anisimov, O., B. Fitzharris, J.O. Hagen, R.L. Jefferies, H. Marchant, F. Nelson, T. Prowse, and D.G. Vaughan. 2001. Polar Regions (Arctic and Antarctic). In J.J. McCarthy, O.F. Canziani, N.A. Leary, D.J. Dokken, and K.S. White (Eds.), *Climate Change 2001: Impacts, Adaptation, and Vulnerability* (pp. 803-41). Cambridge, UK: Cambridge University Press.

Bard, S.M. 1999. Global Transport of Anthropogenic Contaminants and the Consequences for the Arctic Marine Ecosystem. *Marine Pollution Bulletin 38,* 356-79.

Berkes, F. 1999. *Sacred Ecology: Traditional Ecological Knowledge and Resource Management.* Philadelphia: Taylor and Francis.

Berkes, F., N. Bankes, M. Marschke, D. Armitage, and D. Clark. 2005a. Cross-Scale Institutions and Building Resilience in the Canadian North. In F. Berkes, R. Huebert, H. Fast, M. Manseau, and A. Diduck (Eds.), *Breaking Ice: Renewable Resource and Ocean Management in the Canadian North* (pp. 225-47). Calgary: University of Calgary Press.

Berkes, F., R. Huebert, H. Fast, M. Manseau, and A. Diduck (Eds.). 2005b. *Breaking Ice: Renewable Resource and Ocean Management in the Canadian North.* Calgary: University of Calgary Press.

Berkes, F., and D. Jolly. 2001. Adapting to Climate Change: Social-Ecological Resilience in a Canadian Western Arctic Community. *Conservation Ecology, 5,* 18. http://www.consecol.org/vol15/iss12/art18.

Berner, J., and C. Furgal. 2005. Human Health. In ACIA (Ed.), *Impacts of a Warming Climate: Arctic Climate Impact Assessment* (pp. 863-906). Cambridge, UK: Cambridge University Press.

Borga, K., G.W. Gabrielsen, and J.U. Skaare. 2001. Biomagnification of Organochlorines along a Barents Sea Food Chain. *Environmental Pollution, 113,* 187-98.

Buccini, J.A. 2003. The Long and Winding Road to Stockholm: The View from the Chair. In D.L. Downie and T. Fenge (Eds.), *Northern Lights against POPs: Combatting Toxic Threats in the Arctic* (pp. 224-55). Montreal: McGill-Queen's University Press.

Burger, J., M. Gochfeld, C.W. Powers, D.S. Kosson, J. Halverson, G. Siekaniek, A. Morkill, R. Patrick, L.K. Duffy, and D. Barnes. 2006. Scientific Research, Stakeholders, and Policy: Continuing Dialogue during Research on Radionuclides on Amchitka Island, Alaska. *Journal of Environmental Management, 85*(1), 557-68.

Chapin, F.S. III, M. Berman, T.V. Callaghan, P. Convey, A.-S. Crepin, K. Danell, H.W. Ducklow, B. Forbes, G. Kofinas, A.D. McGuire, M. Nuttall, R.A. Virginia, O.R. Young, and S. Zimov. 2005a. Polar Systems. In R. Hassan, R. Scholes, and N. Ash (Eds.), *Millennium Ecosystem Assessment: Ecosystems and Human Well-Being: Current State and Trends* (Vol. 1.) (pp. 717-43). Washington, DC: Island Press.

Chapin, F.S. III, A.L. Lovecraft, E.S. Zavaleta, J. Nelson, M.D. Robards, G.P. Kofinas, S.F. Trainor, G.D. Peterson, H.P. Huntington, and R.L. Naylor. 2006. Policy Strategies to Address Sustainability of Alaskan Boreal Forests in Response to a Directionally Changing Climate. *Proceedings of the National Academy of Sciences of the United States of America, 103,* 16637-43.

Chapin, F.S. III, G. Peterson, F. Berkes, T.V. Callaghn, P. Angelstam, M. Apps, C. Beier, Y. Bergeron, A.-S. Crein, K. Danell, T. Elmqvist, C. Folke, B. Forbes, N. Fresco, G. Juday, J. Niemela, A. Shvidenko, and G. Whiteman. 2004. Resilience and Vulnerability of Northern Regions to Social and Environmental Change. *Ambio, 33,* 344-49.

Chapin, F.S. III, M. Sturm, M.C. Serreze, J.P. McFadden, J.R. Key, A.H. Lloyd, A.D. McGuire, T.S. Rupp, A.H. Lynch, J.P. Schimel, J. Berigner, W.L. Chapman, H.E. Epstein, E.S. Euskirchen, L.D. Hinzman, G. Jia, C.-L. Ping, K.D. Tape, C.D.C. Thompson, D.A. Walker, and J.M. Welker. 2005b. Role of Land-Surface Changes in Arctic Summer Warming. *Science, 310,* 657-60.

Cone, M. 2005. *Silent Snow: The Slow Poisoning of the Arctic.* New York: Grove Atlantic.

Crowley, T.J. 2000. Causes of Climate Change Over the Past 1000 Years. *Science, 289,* 270-77.

Dewailly, E., P. Ayotte, S. Bruneau, C. Laliberte, D. Muir, and R. Norstrom. 1993. Inuit Exposure to Organochlorines through the Aquatic Food Chain in Arctic Quebec. *Environmental Health Perspectives, 101,* 618-20.

Dewailly, E., and C. Furgal. 2003. POPs, the Environment and Public Health. In D.L. Downie and T. Fenge (Eds.), *Northern Lights against POPs: Combating Toxic Threats in the Arctic* (pp. 3-21). Montreal: McGill-Queen's University Press.

Downie, D.L., and T. Fenge (Eds.). 2003. *Northern Lights against POPs: Combatting Toxic Threats in the Arctic.* Montreal: McGill-Queen's University Press.

Dresner, S. 2002. *The Principles of Sustainability.* London: Earthscan Publications.

Duerden, F. 2004. Translating Climate Change Impacts at the Community Level. *Arctic, 57,* 204-12.

Duhaime, G. (Ed.). 2002. *Sustainable Food Security in the Arctic: State of Knowledge.* Edmonton: Canadian Circumpolar Press.

Ebi, K.L., R.S. Kovats, and B. Menne. 2006. An Approach for Assessing Human Health Vulnerability and Public Health Inverventions to Adapt to Climate Change. *Environmental Health Perspectives, 114,* 1930-34.

Faber, D.R., and D. McCarthy. 2003. Neo-liberalism, Globalization and the Struggle for Ecological Democracy: Linking Sustainability and Environmental Justice. In J. Agyeman, R.D. Bullard, and R. Evans (Eds.), *Just Sustainabilities, Development in an Unequal World* (pp. 38-63). Cambridge, MA: MIT Press.

Fenge, T. 2003. POPs and Inuit: Influencing the Global Agenda. In D.L. Downie and T. Fenge (Eds.), *Northern Lights against POPs: Combatting Toxic Threats in the Arctic* (pp. 192-213). Montreal: McGill-Queen's University Press.

Furgal, C.M. 2005. Monitoring as a Community Response for Climate Change and Health. *International Journal of Circumpolar Health, 64,* 440-41.

Furgal, C., and J. Seguin. 2006. Climate Change, Health, and Vulnerability in Canadian Northern Aboriginal Communities. *Environmental Health Perspectives, 114,* 1964-70.

Godduhn, A., and L. Duffy. 2003. Multi-Generation Health Risks of Persistent Organic Pollution in the Far North: Use of the Precautionary Approach in the Stockholm Convention. *Environmental Science and Policy, 6,* 341-53.

Hild, C. 2003. Communicating Risks about Alaska Natives' Food. *Northwest Public Health,* Spring/Summer 2003, 12-13.

–. 1998. Cultural Concerns Regarding Contaminants in Alaskan Local Foods. *International Journal of Circumpolar Health, 57,* 561-66.

Hinzman, L., N. Bettez, F.S. Chapin, M. Dyurgerov, C. Fastie, B. Griffith, R.D. Hollister, A. Hope, H.P. Huntington, A. Jensen, D. Kane, D.R. Klein, G.P. Kofinas, A. Lynch, A. Lloyd, A.D. McGuire, F. Nelson, W.C. Oechel, T. Osterkamp, C. Racine, V. Romanovsky, D. Stow, M. Sturm, C.E. Tweedie, G. Vourlitis, M. Walker, D. Walker, P.J. Webber, J. Welker, K. Winker, and K. Yoshikawa. 2005. Evidence and Implications of Recent Climate Change in Northern Alaska and Other Arctic Regions. *Climate Change, 72,* 251-98.

Huntington, H., and S. Fox. 2005. The Changing Arctic: Indigenous Perspectives. In ACIA (Ed.), *Impacts of a Warming Climate. Arctic Climate Impact Assessment* (pp. 61-98). Cambridge, UK: Cambridge University Press.

Huntington, H.P., S.F. Trainor, D.C. Natcher, O. Huntington, L.O. DeWilde, and F.S.I. Chapin. 2006. The Significance of Context in Community-Based Research: Understanding Discussions about Wildfire in Huslia, Alaska. *Ecology and Society, 11,* 40.

ICC (Inuit Circumpolar Conference). 2005. Executive Council Resolution 2003-01 Regarding Climate Change and Inuit Human Rights. Petition to the Inter-American Commission on Human Rights Seeking Relief from Violations Resulting from Global Warming Caused by Acts and Omissions of the United States. In M.S. Watt-Cloutier (Ed.), *Inuit Circumpolar Conference.* Summary and full petition. http://inuitcircumpolar.com/index.php?auto_slide=&ID=316&Lang=En&Parent_ID=¤t_slide_num=.

Irniq, P. 2004. *Speaking Notes on Climate Change.* Reykjavik: Nunavut Government. http://www.gov.nu.ca/Nunavut/English/departments/commissioner/iceland1.shtml (accessed September 2005).

Jamieson, D. 2001. Climate Change and Global Environmental Justice. In C.A. Miller and P.N. Edwards (Eds.), *Changing the Atmosphere, Expert Knowledge and Environmental Governance* (pp. 287-307). Cambridge, MA: MIT Press.

Jilek, W.G. 1987. The Impact of Alcohol on Small-Scale Societies in the Cirum-Pacific Region. *Curare, 10,* 151-68.

Kinloch, D., H. Kuhnlein, and D. Muir. 1992. Inuit Foods and Diet: A Preliminary Assessment of Benefits and Risks. *Science of the Total Environment, 122,* 247-78.

Kofinas, G., B. Forbes, H. Beach, F. Berkes, M. Berman, T. Chapin, Y. Csonka, D. Kjell, G. Osherenko, T. Semenova, J. Tetlichi, O. Young, and D. Magness. 2005. A Research Plan for the Study of Rapid Change, Resilience and Vulnerability in Social-Ecological Systems in the Arctic. *Common Property Digest, 73,* 1-10.

Kraemer, L.D., J.E. Berner, and C.M. Furgal. 2005. The Potential Impact of Climate on Human Exposure to Contaminants in the Arctic. *International Journal of Circumpolar Health, 64,* 498-508.

Krupnik, I., and D. Jolly (Eds.). 2002. *The Earth Is Faster Now: Indigenous Observations of Arctic Environmental Change.* Fairbanks, AK: Arctic Research Consortium of the United States.

Kurtz, H.E. 2003. Scale Frames and Counter-Scale Frames: Constructing the Problem of Environmental Injustice. *Political Geography, 22,* 887-916.

Leiserowitz, A., R. Gregory, and L. Failing. 2006. *Climate Change Impacts, Vulnerabilities, and Adaptation in Northwest Alaska*. No. 06-11. Eugene, OR: Decision Research.

Lynch, A.H., J.A. Curry, R.D. Brunner, and J.A. Maslanik. 2004. Towards an Integrated Assessment of the Impacts of Extreme Wind Events on Barrow, Alaska. *Bulletin of the American Meteorological Society, 85,* 209-21.

McBean, G., G. Alekseev, D. Chen, E. Forland, J. Fyfe, P.Y. Groisman, R. King, H. Melling, R. Vose, and P.H. Whitefield. 2005. Arctic Climate: Past and Present. In ACIA (Ed.), *Impacts of a Warming Climate: Arctic Climate Impact Assessment* (pp. 22-60). Cambridge, UK: Cambridge University Press.

McCarthy, J., and M.L. Martello. 2005. Climate Change in the Context of Multiple Stressors and Resilience. In ACIA (Ed.), *Impacts of a Warming Climate: Arctic Climate Impact Assessment* (pp. 945-88). Cambridge, UK: Cambridge University Press.

McCreanor, L. 2003. Levels of Persistent Organic Pollutants (POPs) in a Subarctic Aquatic Ecosystem: Implications for Assessment and Remediation of Northern Contaminated Sites (MA thesis, University of Waterloo, Ontario, 2003).

Middaugh, J. 2001. Arctic Contaminants and Human Health. In J. Burger, E. Ostrom, R. Norgaard, D. Policansky, and B. Goldstein (Eds.), *Protecting the Commons: A Framework for Resource Management in the Americas* (pp. 241-52). Washington, DC: Island Press.

Morrow, P. 1996. Yup'ik Eskimo Agents and American Legal Agencies: Perspectives on Compliance and Resistance. *Journal of the Royal Anthropological Institute, 2,* 405-23.

Myers, H., H. Fast, M. Berkes, and F. Berkes. 2005. Feeding the Family in Times of Change. In A.D. Fikret Berkes, H. Fast, R. Huebert, and M. Manseau (Eds.), *Breaking Ice: Integrated Ocean Management in the Canadian North* (pp. 23-45). Calgary: University of Calgary Press.

Myers, H., and C. Furgal. 2006. Long-range Transport of Information: Are Arctic Residents Getting the Message about Contaminants? *Arctic, 59,* 47-60.

National Research Council. 2001. Climate Change Science: An Analysis of Some Key Questions. National Research Council, Committee on the Science of Climate Change, Division on Earth and Life Studies, Washington, DC.

Northern Contaminants Programme. 1997. *Highlights of the Canadian Arctic Contaminants Assessment Report: A Community Reference Manual.* Ottawa: Northern Contaminants Programme, Indian and Northern Affairs Canada.

Nuttall, M. 2001. Indigenous Peoples and Climate Change Research in the Arctic. *Indigenous Affairs, 4,* 26-35.

–. 2000. Indigenous Peoples' Organisations and Arctic Environmental Co-operation. In M. Nuttall and T. Callaghan (Eds.), *The Arctic: Environment, People, Policy* (pp. 621-37). Amsterdam: Harwood Academic Publishers.

O'Brien, K.L., and R.M. Leichenko. 2003. Winners and Losers in the Context of Global Change. *Annuals of the Association of American Geographers 93,* 89-103.

Osherenko, G., and O.R. Young. 1989. *The Age of the Arctic: Hot Conflicts and Cold Realities.* Cambridge, UK: Cambridge University Press.

Overpeck, J.T., M. Sturm, J.A. Francis, D.K. Perovich, M.C. Serreze, R. Benner, E.C. Carmack, and I.F.S. Chapin. 2005. Arctic System on Trajectory to New, Seasonally Ice-free State. *Eos, 86,* 209-21.

Pellow, D.N. 2000. Environmental Inequality Formation: Toward a Theory of Environmental Injustice. *American Behavioral Scientist, 43,* 581-601.

Peters, E. 2002. Sustainable Development, Food Security and Aboriginal Self-Government in the Circumpolar. In G. Duhaime (Ed.), *Sustainable Food Security in the Arctic: State of Knowledge* (pp. 205-25). Edmonton: Canadian Circumpolar Press.

Pinguelli-Rosa, L., and M. Munashinghe (Eds.). 2002. *Ethics, Equity and International Negotiations on Climate Change.* Cheltenham, UK, and Northampton, MA: Edward Elgar.

Reiersen, L.O. 2000. Local and Transboundary Pollution. In M. Nuttall and T. Callaghan (Eds.), *The Arctic: Environment, People, Policy* (pp. 575-99). Amsterdam: Harwood Academic Publishers.

Sachs, J.D., and W.V. Reid. 2006. Investments toward Sustainable Development. *Science, 312,* 1002.

Schultz, W. 2004. The Many Faces of Chlorine. *Chemical and Engineering News, 82,* 40-42.

Serreze, M.C., J.E. Walsh, I.F.S. Chapin, T. Osterkamp, M. Dyurgerov, V. Romanovsky, W.C. Oechel, J. Morison, T. Zhang, and R.G. Barry. 2000. Observational Evidence of Recent Change in the Northern High-Latitude Environment. *Climatic Change, 46,* 159-207.

Smit, B., and O. Pilifosova. 2001. Adaptation to Climate Change in the Context of Sustainable Development and Equity. In J.J. McCarthy, O.F. Canziani, N.A. Leary, D.J. Dokken, and K.S. White (Eds.), *Climate Change 2001: Impacts, Adaptation, and Vulnerability – Contribution of Working Group II to the Third Assessment Report of the Intergovernmental Panel on Climate Change* (pp. 879-906). Cambridge, UK: Cambridge University Press.

Suedel, B., J. Boraczek, R. Peddicord, P. Clifford, and T. Dillon. 1994. Trophic Transfer and Biomagnification Potential of Contaminants in Aquatic Ecosystems. *Reviews of Environmental Contamination and Toxicology, 136,* 21-89.

Thompson, S. 2005. Sustainability and Vulnerability: Aboriginal Arctic Food Security in a Toxic World. In A.D. Fikret Berkes, H. Fast, R. Huebert, and M. Manseau (Eds.), *Breaking Ice: Integrated Ocean Management in the Canadian North* (pp. 47-69). Calgary: University of Calgary Press.

Thornton, J. 2000. *Pandora's Poison: Chlorine, Health, and a New Environmental Strategy.* Cambridge, MA: MIT Press.

Tyrrell, M. 2006. Making Sense of Contaminants: A Case Study of Arviat, Nunavut. *Arctic, 59,* 370-80.

Van Oostdam, J., S. Donaldson, M. Feeley, D. Arnold, P. Ayotte, G. Bondy, L. Chan, E. Dewailly, C. Furgal, H. Kuhnlein, E. Loring, G. Muckle, E. Myles, O. Receveur, B. Tracy, U. Gill, and S. Kalhok. 2005. Human Health Implications of Environmental Contaminants in Arctic Canada: A Review. *Science of the Total Environment, 351-352,* 165-246.

Van Oostdam, J., A. Gilman, E. Dewailly, P. Usher, B. Wheatley, H. Kuhnlein, S. Neve, J. Walker, B. Tracy, M. Feeley, V. Jerome, and B. Kwavnick. 1999. Human Health Implications of Environmental Contaminants in Arctic Canada: A Review. *Science of the Total Environment, 230,* 1-82.

Watt-Cloutier, S. 2003. The Inuit Journey toward a POPs-Free World. In D.L. Downie and T. Fenge (Eds.), *Northern Lights against POPs: Combatting Toxic Threats in the Arctic* (pp. 256-67). Montreal and Kingston: McGill-Queen's University Press.

Watt-Cloutier, S., T. Fenge, and P. Crowley. n.d. *Responding to Global Climate Change: The Perspective of the Inuit Circumpolar Conference on the Arctic Climate Impact Assessment.* http://inuitcircumpolar.com/index.php?ID=267&Lang=En.

Weiss, K.R. 2008, 15 May. Polar Bear Is Listed as Threatened Species. *Los Angeles Times.*

Williams, R.W. 1999. Environmental Injustice in America and Its Politics of Scale. *Political Geography, 18,* 49-73.

Young, O.R. 2005. Governing the Arctic: From Cold War Theater to Mosaic of Cooperation. *Global Governance, 11,* 9-15.

9
Environmental Justice and Community-Based Ecosystem Management

Maureen G. Reed

> There is, of course, a geography to the process of preservation, and it is uneven.
>
> – C. Katz,
> "Whose Nature, Whose Culture," p. 49

> The problem of conservation cannot be separated from problems of governance and of equity.
>
> – D. Ludwig,
> "The Era of Management Is Over," p. 759

In a recent and cogent reflection on community-based ecosystem management, Berkes (2004, p. 622) identifies three conceptual shifts in applied ecology (broadly defined), including "a shift from reductionism to a systems view of the world, a shift to include humans in the ecosystem, and a shift from an expert-based approach to participatory conservation and management." Indeed, some advocates of community-based ecosystem management argue that social equity can be achieved through collaborative decision-making processes that integrate scientific, environmental, corporate, and community interests with public agencies (e.g., Daniels and Walker, 1996; McLain and Lee, 1996; Cortner and Moote, 1999). Yet, as Cindy Katz (1998) points out, the geography of environmental care is an uneven one. Local capacity and efficacy are unevenly distributed across the landscape, while different landscapes garner uneven levels of concern from nonlocal actors. Consequently, with its emphasis on the creation of local programs tailored to local ecological and social conditions, community-based ecosystem management may contribute to highly differentiated application of conservation practices. Donald Ludwig (2001) provides a hint of how to examine this issue by pointing out the central importance of governance arrangements. Thus, the purpose of this chapter is to explore institutional processes that give rise to uneven environmental protection and to then consider the implications from the perspective of environmental justice.

I examine practices in two Canadian biosphere reserves: Redberry Lake, Saskatchewan, and Clayoquot Sound, British Columbia. I begin by describing connections between spatial and environmental justice and argue that comparative analysis can help to determine a link between unevenness and (in)equity. Next, I provide some background to the concepts of biosphere reserves and environmental management regimes, and I describe how I conducted the research. In two analytical sections, I discuss the emergence of a private stewardship regime at Redberry Lake and a more public regime at Clayoquot Sound. From this analysis, I draw conclusions about the emergence of an uneven level of environmental care and some implications for the biosphere reserves and for our understanding of environmental management processes.

Environmental Justice and Uneven Environmental Management

Advocates and scholars of environmental justice confront broad issues of social inequality based on land-use planning, human settlement, and industrial development, and seek to link concerns for a range of environmental and social causes, including environmental quality, poverty, racism, sexism, and democracy. Bruce Mitchell (2002) defines environmental justice as a set of movements and research programs designed to secure "the right to a safe, healthy, productive and sustainable environment for all" (p. 557). These include paying attention to "the conditions in which such a right can be freely exercised, through which individual and group identities, needs, and dignities are preserved, fulfilled, and respected in a way that provides for self-actualization and personal and community empowerment" (p. 557). David Schlosberg (2004) expands on this definition by arguing for the presence of three components: distribution, recognition, and participation. He states that environmental justice requires equity in the distribution of environmental risk, as well as recognition of the diversity of participants and experiences in affected communities. Only with recognition and a commitment to equity can there be adequate participation in the political processes that create and manage environmental policy (p. 517).

Consequently, environmental justice is concerned with recognition of those affected by environmental policy or change, fair processes to involve them, and equitable outcomes of environmental decisions over time, over space, and across different social groups. Thus, geographic or spatial justice has historically referred to the location and spatial configuration of communities and their proximity to environmental hazards or undesirable land uses, while social justice refers to the ways in which sociological and demographic factors such as race, ethnicity, class, culture, lifestyles, and political power influence environmental decision making. Emphasis by researchers and activists has been on protection from harm. Yet one might consider the relative efficacy of social and geographic groups to engage in environmental stewardship and conservation practices that enhance their local environments, rather than simply reduce their exposure to harm (see also Clapp, 2004).

Geographers have long explored the factors giving rise to and the implications of uneven development (Smith, 1984). Geographers have explored implications of uneven development for land and resource allocation and access (e.g., Lowe, Murdoch, Marsden, Munton, and Flynn, 1993; Roberts and Emel, 1992) and environmental politics (e.g., Pulido, 1996). However, comparative research that links unevenness of environmental management practices and social equity is limited. Although spatial unevenness of environmental management practices do not necessarily lead to social inequality, in rural areas of Canada questions of equity arise where environmental management may demand that resource users allocate lands for new purposes, engage in new management practices, and bear the costs of transition to new forms of production and/or consumption. For example, there is considerable literature on the influence of new, affluent residents who seek to alter the patterns of economic activity in areas of traditional agriculture or resource extraction (e.g., Lowe et al., 1993; Brogden and Greenberg, 2003). Institutions such as law, property regimes, planning, regulation, and markets are produced through political processes that are geographically uneven in form and scale. These variable institutions can give rise to or reinforce processes or outcomes that have differential and sometimes adverse effects on both local livelihoods and biodiversity.

For example, research dealing with conversion of lands for consumptive (e.g., ecotourism) rather than productive (e.g., resource extraction) purposes suggests a heightened role for nongovernmental actors in influencing the land allocation and management practices at the local level, claiming that nongovernmental organizations or interests can gain a relative advantage in creating the environmental and recreational landscapes they desire because of their ability to harness strategic, logistical, or financial resources (Lowe et al., 1993; Ward, Lowe, Seymour, and Clark, 1995; Reed, 1997). Similarly, political economists and political ecologists studying uneven environmental management practices in North America illustrate how institutions of property exchange, reterritorialization, valuation, and planning contribute to or reinforce social inequities as some groups gain power over others in the decision making and in environmental management practices (Heynen and Robbins, 2005; Hurley and Walker, 2004; Robertson, 2005; Lane, 2003; Brogden and Greenberg, 2003). Such institutions affect decisions about who becomes recognized as a stakeholder and how outcomes are distributed; they also affect the ability and efficacy of participants to achieve just outcomes.

Very little of this work is explicitly comparative. Examining cases simultaneously can help to expose commonalities and distinctiveness between local situations, to identify key variables, and to illustrate how they interact. Comparative work can therefore help to understand *how* and *why* particular social and environmental circumstances give rise to particular local outcomes. Comparison, then, can help to determine whether there is a link between unevenness and inequity.

Background to the Study

My research focused on describing the emergence of environmental management regimes that include formal and informal institutional arrangements in which public, private, and civic interests work simultaneously (together or apart, in sync or at odds with one another) to influence, make, and/or carry out governing decisions about environmental and resource management (Reed, 2007). Ecosystem management is but one form of environmental management regime. Because institutional arrangements change over time, environmental management regimes are not uniform or stable across time or space. They take on diverse forms and change character depending on ecological conditions, shifting interests and alliances, available logistical resources and activities, and management efforts of local, provincial, and federal governments. They may be characterized by conflict or by consensual management practices. Public, private, or civic groups may organize to influence and even deliver a policy or program; their efforts may be enhanced or limited by the human, social, cultural, and economic assets of their memberships and their strategic location within other competing political and economic agendas. Furthermore, environmental management regimes are influenced by changing economies and environmental circumstances, shifting socio-cultural relations, as well as by governance practices and institutional capacities across scales. These elements alter the parameters for governance, defined here as institutional arrangements that extend beyond government to include private sector and other nongovernmental organizations, as well as the rule systems under which these different actors operate (Francis, 2003). This perspective about governance raises questions about the relative influence of public and private interests, and scientific and local knowledge in setting and pursuing priorities for conservation.

My research was undertaken in two Canadian biosphere reserves. Biosphere reserves are geographic areas designated because of the expressed desire of local communities to work toward sustainability. Residents seeking biosphere reserve status for their region must have it nominated at the local level, endorsed by provincial and national governments, and, finally, recognized by the United Nations Education, Scientific and Cultural Organization (UNESCO). All biosphere reserves are created to demonstrate three functions: environmental protection, logistical provisioning for scientific research and education, and sustainable resource use. Biosphere reserves contain three zones: a core that must be a protected by legislation, a buffer where research and recreational uses compatible with ecological protection are allowed, and a transition zone (or zone of cooperation) where sustainable resource use is practiced (UNESCO, 2000).

Canadian biosphere reserves are considered protected areas, falling between categories 2 and 6 of the classification scheme of the World Conservation Union. In Canada, the core areas of biosphere reserves are often, although not always, national parks which are typically classified within category 2. Once created, biosphere reserves are managed by local community committees. In this context, management means that local committees obtain funds to undertake educational and demonstration projects and provide logistical support for scientific research. In Canada, at least, these committees do not have regulatory authority or direct management and decision-making powers but must operate within provincial and federal legislative frameworks and/or work with relevant government agencies in cooperative decision-making forums.

The research reported here was conducted at Redberry Lake and Clayoquot Sound Biosphere Reserves between 2001 and 2004 (see Figure 9.1). During this time, my research assistant and I spent about six weeks in each locality. We attended meetings of the biosphere reserve committees and local events; obtained and reviewed minutes of previous meetings, records, and personal notes and letters; and analyzed official nomination papers and related government documents. In addition, we conducted interviews with all the active members of both biosphere reserve committees, as well as with government workers and environmental nongovernmental organizations in each locality that had direct involvement with the reserves. Forty-four semi-formal interviews were tape-recorded and transcribed verbatim for analysis. They followed a consistent interview schedule under topical areas: involvement in the biosphere reserve, land ownership and use, livelihood strategies and local economy, participatory decision making and partnership arrangements, and socio-demographic characteristics. Thirteen further interviews followed a less formal schedule but were also tape-recorded and transcribed verbatim for analysis. I undertook analysis by both deductive coding (predetermined from within the above categories) and inductive coding (emerging from interviews themselves) using ATLAS.ti, a qualitative data analysis software program. I aggregated coded

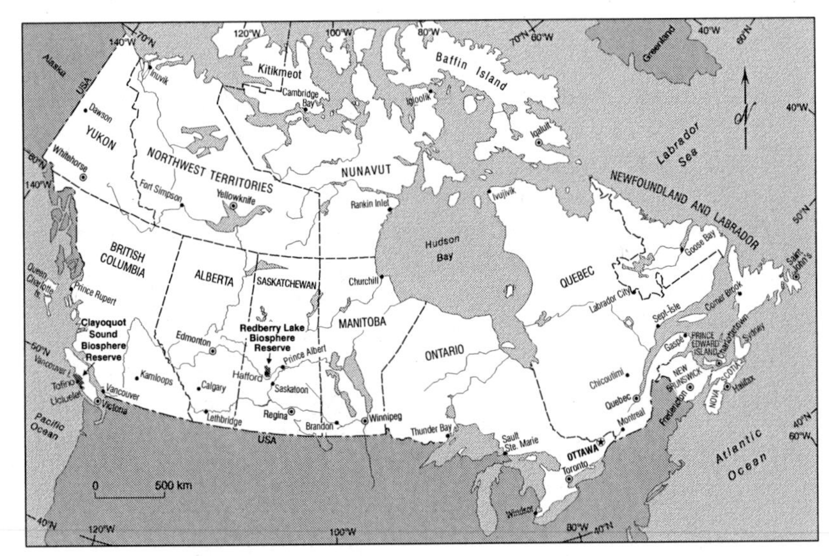

Figure 9.1 Location of the Clayoquot Sound and Redberry Lake biosphere reserves in Canada. Map by Keith Bigelow.

data into themes and verified the accuracy of individual observations against other interviews and documented sources (after Cresswell, 1998). I also prepared a draft paper for review by key informants in each locality. Between 2004 and 2006, I ceased active research in the biosphere reserves, although I maintained contact with local people and have remained current with key activities in each region.

Redberry Lake: The Emergence of Private Stewardship

Located in the central Canadian Prairie province of Saskatchewan, Redberry Lake was designated a biosphere reserve by UNESCO in 2000. Much of the biosphere reserve's ecological importance rests with the aquatic environment that supports waterfowl and shorebird populations that are considered globally and nationally significant (Schmutz, 1999). As part of the Aspen Parkland region, many habitats and wildlife that would have been present at Redberry were extirpated at the time of the fur traders and settlement. Thus, by the late nineteenth and early twentieth centuries, the region was already a highly modified landscape (McGillivray, 1998). Continuous regional modification is a result of settlement, agricultural production, and industrialization.

The core of the Redberry Lake Biosphere Reserve is the lake itself, protected by the federal Migratory Birds Convention Act, with islands protected as provincial wildlife reserves. The core area composes only five percent of the biosphere reserve. The buffer zone of the Redberry Lake

Biosphere Reserve is also relatively small, composing only 6,300 hectares immediately surrounding the lake, or 6 percent of the total reserve area. Of this, 920 hectares are protected under the provincial Wildlife Habitat Protection Act. A regional park was created for the remainder to ensure that land developments at the lakefront would not jeopardize habitat provided for waterfowl and shorebirds. The transition zone, where the town of Hafford and the rural municipality of Redberry are located, composes 89 percent of the land area of the biosphere reserve. The dominant resource activity is agriculture and, therefore, most of the land in the transition zone of the biosphere reserve is privately owned.

According to hydrographic data, the saline lake for which Redberry is named has dropped 4.15 metres in the past thirty-seven years (Saskatchewan Watershed Authority, 2003), reducing the habitat for shorebirds that use the lake for nesting or staging. The most significant drop in lake level occurred from 2000 to 2003. The buffer zone of the biosphere offers little protection from the changes in the lake itself or from the agricultural inputs from nearby farms. Few members of the public know about the status of the lake, and few federal officers come to enforce the legislation. Beginning in 1989, the Canadian Wildlife Service (CWS), a federal agency, contracted the Redberry Pelican Project to monitor the effects of visitors on nest disturbance, undertake regular patrols of the bird sanctuary, and report violations to authorities. The group also provided public awareness and education to local residents through an interpretive centre built by the lake in 1992. During the 1990s, funding from the CWS declined and by 2001 lake monitoring was terminated (Anonymous, 2002).

At Redberry Lake, there are very few public instruments for environmental conservation, and relatively few environmental movement organizations present in the region. Most land is held privately, and legislation regulates individual users rather than regional priorities. Acts and regulations established for agriculture in Saskatchewan tend to focus on operational issues of individual properties (e.g., Environmental Farm Plans), with issues such as nutrient management and integrated pest management identified as key components of the shift toward more sustainable agriculture. One 2006 initiative, the Canada-Saskatchewan Farm Stewardship Plan, encourages landscape or regional-scale planning within a geographic area defined by physical boundaries (e.g., a watershed), to address local agri-environmental issues. Upon completion of the group plan, producer members of the group are eligible to apply for financial incentives to assist them in implementing best management practices for the issues they have chosen.

Despite such a government program, the general lack of government interest in the Redberry region over the course of the study period was striking, particularly given the necessity of provincial and federal endorsement of the UNESCO nomination. Although civil servants working out of the provincial capital of Regina received funding requests by Redberry Lake Biosphere Reserve members since 2000, neither the provincial and nor the federal levels of government were directly involved with the biosphere reserve on a regular basis. In addition, the biosphere reserve committee has not been successful in securing any Aboriginal involvement, although there is a nonresident First Nations reserve within the biosphere reserve region.

Until 2006, nine regular volunteers, of whom six are board members, served the Redberry Lake Biosphere Reserve committee. Upon designation, the committee contracted a consultant to engage in community meetings and develop a sustainability plan (Sian, 2001). However, few of the recommendations of the plan have been implemented. Since its inception, the committee made annual requests to the provincial government to provide core funding to maintain research, education, and public information activities and to keep its interpretive centre in operation. Until 2005, these requests met with very modest levels of funding, provided to the biosphere reserve on an annual basis. In 2005, five-year core funding was specified in the provincial budget and thereby opportunities for new initiatives were created.[1]

A general lack of government interest has been mirrored by environmental nongovernmental organizations (ENGOs). The region has not attracted strong interest from other provincial ENGOs, such as the Saskatchewan Environmental Society or Nature Saskatchewan. The lack of collaboration between the biosphere reserve committee and other ENGOs, the dominance of private property in the region, and the lack of sustained interest from the provincial government has opened the door for an environmental management regime based on privately driven institutional arrangements. Two national environmental groups with conservation mandates were active in the Redberry region at the time of designation: Ducks Unlimited Canada (DUC) and the Nature Conservancy of Canada (NCC). These organizations have undertaken important conservation work within and outside the biosphere reserve by working together and with landowners to protect habitat. Joint ownership, land donations, stewardship agreements, and conservation easements are all mechanisms for ENGOs to place land under a protective management regime (Ducks Unlimited Canada, 2004a; Nature Conservancy of Canada, n.d.). Both the NCC and DUC establish their own national conservation priorities before negotiating with individual landowners or acquiring land for stewardship values rather than for traditional extractive practices. There is no effort to work with local groups to collectively establish priorities or management options.

Both organizations place a high value on the "best available conservation science" (Nature Conservancy of Canada, 2001; see also Ducks Unlimited

Canada, 2004b) to identify priority landscapes for their activities. In promotional materials, NCC boasts of making use of the best available conservation science (Nature Conservancy of Canada, 2001), while DUC refers to the Institute of Wetland and Waterfowl Research as its "highly respected scientific research arm" (Ducks Unlimited Canada, 2004b). Neither organization refers to traditional or local ecological knowledge as informing their planning or initiatives at the landscape level. Rather, they each refer to the adjudication of their respective plans by "scientific experts." Neither organization lists accountability to local communities as key elements of their agendas. Rather, efforts focus first on engaging with government and academic scientists to identify key elements, as well as expert review once a plan is near completion (interviews with government and ENGO representatives, by author, 2002). Subsequent efforts are directed to securing properties or establishing other conservation programs in each region. Thus, while local landowners may be part of the execution of these plans, there are no direct means of involving local residents in the key planning and decision-making phases (Nature Conservancy of Canada, 2003). Although these activities are not strictly confidential, they are not widely publicized within the affected localities. Even the newsletter is not publicly available. The link with science is viewed not merely as unproblematic by these organizations but as appropriate justification for planning and practices that result.

For some time, DUC has worked quietly on a range of conservation initiatives in the Redberry region. DUC projects include habitat purchases that are cost-shared with monies provided by the provincial ministry, Saskatchewan Environment, as well as a project related to riparian habitat protection that it has shared with NCC. In addition, it has shared ecological information requested by scientific researchers working in the area. When interviewed, the DUC representative revealed that he was very knowledgeable about the biosphere reserve and identified several common goals based on his reading of the community plan, including the development of educational programming, water-quality monitoring, and sustainable agriculture. Beyond these stated common goals, the DUC interviewee identified initiatives that could be shared, including monitoring loss of wetland and native/natural upland habitats, and promoting understanding about how human uses in the biosphere reserve impact its ecological integrity.

These two organizations are not averse to working with others. For example, NCC and DUC worked together to protect riparian habitat at Oscar Creek within the core area. Yet despite their shared interests in principle, neither the NCC nor DUC was able to connect with the Redberry Lake Biosphere Reserve committee. Proposals on the part of the ENGOs were greeted with a sense of distrust by members of the biosphere reserve committee, who were reluctant to establish partnerships with external organizations that would not provide any financial benefits. According to one board member who is appointed by the regional municipality, some biosphere reserve committee members had a fear of opening the door to national partnerships for fear of losing local control.

Some members of the community also expressed concern that national ENGOs negotiated land donations, stewardship agreements, or conservation easements with individual landowners without the knowledge of the adjacent owners. These activities were viewed as competitive with traditional uses and exclusive because other groups were effectively excluded from the governance (i.e., the planning and decision-making phases) of conservation. That is, local residents were not recognized as legitimate stakeholders unless they were landowners with property of ecological significance (as determined by external processes). Other interviewees expressed that philosophies, goals, and management approaches of these two national ENGOs were in marked contrast to interviewees' interpretations of the broader Prairie cultural tradition that focuses on the importance of the small producer and *collective* goodwill. Consequently, the external ENGOs found it was more effective to work on their own. Thus, through mutual distrust and claims of exclusion, partnerships for more effective governance of the biosphere reserve as a whole failed to become established.

In the particular case of Redberry Lake, both the tools that have been used (property instruments) and the ways of wielding them (in confidential negotiations and closed planning processes) became more private than public in character. From an ecological perspective, the presence of NCC and DUC were key to the protection of specific habitats. Without these specific organizations, the detailed projects would not have been undertaken. But from a social perspective, critical elements of environmental management practice (including assessing ecological significance of particular properties and regions, determining the required level and type of intervention, composing and implementing enforcement and monitoring strategies) were initiated by ENGOs in the absence of broader community participation. Thus, initiatives by ENGOS achieved an improved distribution of protected habitats for bird populations in the area, but they did not recognize local producers or engage them in political processes to identify ecologically significant areas or consider the best ways to protect them. The gaps in recognition and political process are even more striking when the situation at Redberry Lake is compared with that of Clayoquot Sound.

Clayoquot Sound: Establishing a Robust – If Acrimonious – Public Regime

Differences in spatial extent, economic activity, and property regimes in British Columbia influenced the emergence of quite a different kind of the environmental management regime in Clayoquot Sound. The Clayoquot

Sound region, on Canada's west coast, has gained international and national recognition as a spectacular example of unlogged temperate rainforest that includes marine, coastal, and terrestrial ecosystems of high biodiversity and cultural values (for a description, see Magnusson and Shaw, 2003). Clayoquot Sound is also located in a region of high population growth, attracting people of all ages who have high levels of formal education. In Clayoquot, the core area is protected by the Pacific Rim National Park Reserve, established under the Canada National Parks Act. The act also has several associated regulations and planning mechanisms to reinforce its ecological and cultural-integrity policies. These are administered locally by the equivalent of a full-time staff of thirty-three in winter and seventy-nine in summer (interview by the author). Notwithstanding budget cuts to Parks Canada during the 1990s (see Dearden and Dempsey, 2004) and increased visitation pressures, Pacific Rim National Park enjoys a staffing presence capable of planning, executing, and enforcing policies and plans that Redberry Lake lacks.

Local and extra-local environmental organizations have continued to hold government – particularly at provincial and federal levels – accountable for environmental protection. Governments at all levels, including municipal and First Nations, along with environmental nongovernmental organizations, actively participate in local processes. Support for protecting this landscape is provided not only by links to international organizations but also by a growing local population with high levels of formal education to which government officials are likely to respond because they share similar cultural and social characteristics (Young, 1990; Halvorsen, 2001). Because of their socio-economic status and their links with outside organizations, the residents also have the ability to apply a credible threat to operations that do not meet environmental standards by garnering international attention and withdrawing support. The result is a wide array of private firms and not-for-profit organizations that are active in shaping the regional environment. They work separately and together to ensure that the governing institutions remain open and public in character, and locally based organizations (e.g., Friends of Clayoquot Sound, the Clayoquot Sound Central Region Board) are in a position to make government initiatives and university-based scientific research accountable. At Clayoquot Sound, a more robust – if sometimes acrimonious – and public system of governance has emerged.

Property rights in the region reinforce a public environmental management regime. At Clayoquot, core and buffer areas make up 49 percent of the biosphere reserve, with only 51 percent of the land area in the transition zone. This land is mostly provincial Crown (public) land where forestry is the dominant activity. Forestry is primarily undertaken on public lands, and forestry regulations have long guided the management of a wide range of ecosystem components (e.g., fish, water quality, soils, and vegetation). Specific regulations are buttressed by public land-use planning initiatives dating back thirty years, but most recently shaped by the influential Clayoquot Sound Scientific Panel for Sustainable Forest Practices (1994, 1995).

In Clayoquot Sound, scientific research has begun to bridge local and conventional scientific knowledge. For example, the government-appointed scientists who formed the Clayoquot Sound Scientific Panel for Sustainable Forest Practices in the mid-1990s argued for changes in logging practices in the region on the basis of the best available science *and* the best possible traditional ecological knowledge of First Peoples of the region. Its reports brought indigenous concepts of ecology to bear directly in its recommendations to the provincial government (e.g., Clayoquot Sound Scientific Panel for Sustainable Forest Practices, 1994, 1995). More recently, a research alliance established in 2001 between the University of Victoria and the Clayoquot Sound Biosphere Reserve spent almost two years developing protocols to ensure that local people, particularly Aboriginal people, would be directly involved in setting the research agenda for the region (see Indigenous Governance Program, 2003). Regular local conferences brought academic researchers (scientists and social scientists) together with local people to discuss research results and needs. Here, science became a publicly vetted activity, rather than an assumed "good."

Thus, the combination of legislation, planning initiatives, and agreements in Clayoquot Sound provides a broad range of instruments to maintain an active but public governance regime. Aboriginal peoples, while diverse in their political interests, were active across all components of the regional resource economy, research, management, and planning activities.

The Clayoquot Sound biosphere reserve is supported by an ongoing funding arrangement. After years of regional conflict over clear-cut logging that became international in scope, both provincial and federal governments had to be seen to take action. In an unprecedented move, the federal government created a trust fund of $12 million. The investment earnings were to be used for ongoing management. Despite initial losses of capital because of instabilities in financial markets and initial local infighting about the best use of the funds, the Clayoquot Biosphere Trust was established and a board of directors created. Although the board is voluntary, the fund provides for a paid executive director. Under his leadership, the reserve area underwent several visioning exercises and a small number of projects began in 2003.[2] By June 2004, three calls for local projects had been made. More projects have been established in subsequent years. The relatively consistent level of funding, along with a functioning board and a full-time, well-paid staff person, have given the biosphere reserve an increased point of leverage to work with other nongovernmental organizations, academics, and local and Aboriginal governments to identify and address local needs.

In the case of Clayoquot Sound, broad participation in the work of the biosphere reserve arose from recognition of the interests and entitlements of local residents, different levels of government, Aboriginal peoples, and private industries in the region. Recognition of such a broad base of interest has been challenging to manage but has given rise to a political process in which ENGOs, among other actor groups, must work openly and often jointly to achieve conservation objectives. The larger number of locally generated projects in Clayoquot Sound compared with Redberry Lake suggests that the distribution of ecological protection efforts also favours Clayoquot. In sum, Clayoquot Sound demonstrated higher levels of distribution of environmental outcomes, greater recognition of diversity within the region, and more inclusive political processes – all elements that favour environmental justice.

Conclusion: Setting the Terms of Environmental Endearment

Clayoquot Sound reveals higher levels of state involvement *and* greater influence by ENGOs operating locally and extra-locally than at Redberry Lake. The severing of ties between conservation organizations at Redberry Lake (such as the biosphere reserve committee, NCC, and DUC), coupled with private property regimes, lack of government oversight, and neglect of the region by other ENGOs, established an environmental management regime at Redberry Lake that was more privately driven than at Clayoquot Sound. Both the tools that were used (property instruments) and the ways of wielding them (in confidential negotiations and planning instruments) were more private than public in character. Consequently, the range of people that could be recognized and involved in political processes was reduced.

The challenge for environmental justice at Redberry Lake is that decisions about environmental protection could effectively be made by a small number of organizations that were not publicly accountable. These decisions may relate to who assesses the ecological significance of particular properties and regions (and how assessments are conducted), who determines the required level of care, who considers how scientific information is collected and applied, who establishes what environmental management practices should be undertaken, and who determines how results should be measured and how management practices are to be enforced.

The small number of people involved at Redberry Lake has fuelled concerns about community disempowerment and about the diminishment of local knowledge versus scientific knowledge. In this case, ongoing planning based on the so-called best available conservation science (Nature Conservancy of Canada, 2001; Ducks Unlimited Canada, 2004b) that will only *later* be brought to public attention reinforces local exclusion and effectively creates a paternalistic style of management, in this case, by an ENGO rather than by government. As NCC and DUC and the biosphere reserve committee shut their doors to one another they risk reducing the collective commitment to conservation of ecosystems and livelihoods shared within the reserve area. Indeed, given the previous federal withdrawal from the monitoring program, the health of the habitat for the bird populations that have been described as "globally and nationally significant" (Schmutz, 1999) remain in the hands of only two ENGOs.

These findings lead to the conclusion that where the public sector is weak, NGOs can (and possibly will) operate more like private organizations that negotiate private contracts than like public or civic organizations. Although unevenness is not synonymous with inequity, practices converge to make the two outcomes at least coincident. These activities reinforce previous research across many contexts that suggests that when governance structures and processes develop unevenly, and in the absence of a sophisticated understanding of power relations, planning processes – no matter how well intentioned – typically reinforce exclusion rather than equity (Berkes, 2004; Reed, 2003).

Comparing the situation in the two biosphere reserves reveals the potential of a widening gap in the provision of environmental management and services whereby biodiversity and habitat protection and conservation practices may become subject to a two-tier system of environmental care. Where local *community* capacity to harness logistical resources, engage in networking, use local knowledge effectively in public forums, and raise funds is weak, at least two possible actions arise. Either no effective action is taken locally or local action relies on private or individual initiative. Both options violate the basic premise of community-based ecosystem management by restricting who gets recognized and who participates and thereby reduces the efficacy of the regime to achieve social and environmental justice.

The conclusion, however, is not to suggest that NGOs such as NCC and DUC move out of environmental management activities. Indeed, their expertise, experience, and conservation commitment are required in the region. Rather, there is a need to assess the potential for strategic combinations of private-public-civic governance arrangements and to strengthen the capacity for organizations to work together instead of relying on singular options and isolated organizations to provide environmental management. Biosphere reserve committees can play a more active role in brokering these arrangements and must find ways to create bridges and links rather than severing ties among organizations with similar overall objectives.

As Schlosberg (2004) points out, environmental justice requires a broader range of participants and avenues for participation. This requirement demands that mechanisms for inclusion and participation be carefully assessed (Reed and McIlveen, 2006) and that improved links be created across spatial and temporal scales. This conclusion is consistent with Barrett, Brandon, Gibson, and Gjertsen's suggestion (2001) that biodiversity protection may best be served by "distributing authority across multiple institutions, rather than concentrating it in just one" (p. 497), as well as with Berkes' call (2004) for cross-scale networks of conservation practices.

In his commentary about community-based conservation, Berkes (2004) recommends that community-based conservation share "management power and responsibility – as opposed to token consultation and passive participation – and [create] a context that encourages learning and stewardship and builds mutual trust" (p. 629). Furthermore, he argued that equity and empowerment requires decision-making processes that are "legitimate, accountable, and inclusive and that take into account multiple stakeholders and interests" (p. 629). These are not idealized recommendations but ones based on years of working with groups around the world. The recommendations, although directed toward international NGOs or state agencies in developing countries, are also relevant in Westernized contexts where both NGOs and state agencies may have quite uneven abilities to influence management regimes.

To date, there has not been much debate over whether there is a universal entitlement in Canada to environmental care across geographic regions and over time (for present and future generations). Yet thinking about environmental justice across such a large and diverse country raises many questions. Do regions share a base entitlement to standards of environmental care? If so, what does this mean operationally when ecosystems (that include biophysical and social components) vary across the country? What should be the relative roles of the public, private, and civic sectors in establishing the associated environmental management regimes? By what criteria should representation across and within those groups be established? What forms of participation will improve the openness and efficacy of decision-making processes (see also Reed and McIlveen, 2006)? Particularly in Western contexts, there is an unstated assumption that the state is universally well organized and that civic organizations have strong commitments and voices in environmental protection debates. Although most advocates of sustainability would likely suggest that basic environmental protections should be granted to all people over time and space, the relative cost of providing these services, management and enforcement capabilities, and local capacities to maintain these protections are highly variable. The implications of this variability for environmental protection and social equity have not been part of serious discussions. In a country as large and geographically, socially, and politically diverse as Canada, where provinces and regions exhibit vastly different levels of commitment and financial capability, these differences are hardly trivial.

The success of environmental management regimes to protect biodiversity and livelihoods will lie in nurturing the well-being of people who must be directly involved in and benefit from environmental protection. This is as true in western Canada as it is in developing nations. Any moves to implement options should be designed to help rural people by assisting them to publicly debate and organize their own priorities among production and stewardship goals; invest in environmental recovery, rehabilitation, and ecological integrity; maintain environmental and amenity values for the broader public; and respect the cultural values of Aboriginal and settler populations. Only by doing so can we promote social well-being and environmental justice for present and future generations. Strengthening institutional arrangements that recognize groups and improve the efficacy of their participation will improve the distribution of environmental protection initiatives, assist in making conservation practice locally inclusive, and ultimately, move toward attaining environmental justice.

Notes

1 In 2005, the biosphere reserve succeeded in obtaining core funding from the provincial government for five years. With this funding, the reserve renovated its original interpretive centre, reopening it in 2006 as a research and education centre. New connections with the local school were established, and the school was granted a national award for its work in environmental education and activism. The biosphere reserve committee publicly stated that it sought new partnerships with organizations working in the region, including the Nature Conservancy of Canada and Ducks Unlimited Canada. Further study and analysis will reveal if such statements are followed up by specific actions.

2 Interviews, letters, and meeting notes reveal that the early days of the biosphere and the trust were stormy ones. In 2002, a new executive director was appointed who was very successful in bringing previously antagonistic parties together for discussion to set priorities and initiate tangible projects.

References

Anonymous. 2002. 15 Years: 15 Years towards Sustainable Development at the Redberry Lake Biosphere Reserve, 1998-2002. Unpublished report, Hafford, Redberry Lake Biosphere Reserve.

Barrett, C.B., K. Brandon, G. Gibson, and H. Gjertsen. 2001. Conserving Tropical Biodiversity amid Weak Institutions. *Bioscience, 51,* 497-502.

Berkes F. 2004. Rethinking Community-Based Conservation. *Conservation Biology, 18,* 621-30.

Brogden M.J., and J.B. Greenberg. 2003. The Fight for the West: A Political Ecology of Land Use Conflicts in Arizona. *Human Organization, 62,* 289-98.

Clapp, A. 2004. Wilderness Ethics and Political Ecology: Remapping the Great Bear Rainforest. *Political Geography, 23,* 839-62.

Clayoquot Sound Scientific Panel for Sustainable Forest Practices. 1995. *Report 3: First Nations Perspectives Relating to Forest Practices Standards in Clayoquot Sound.* Victoria, BC: Queen's Printer.

–. 1994. *Report 1: Report of the Scientific Panel for Sustainable Forest Practices in Clayoquot Sound*. Victoria, BC: Queen's Printer.
Cortner, H.J., and M.A. Moote. 1999. *The Politics of Ecosystem Management*. Washington, DC: Island Press.
Cresswell, J. 1998. *Qualitative Inquiry and Research Design*. Thousand Oaks, CA: Sage.
Daniels, S.E., and G.B. Walker. 1996. Collaborative Learning: Improving Public Deliberation in Ecosystem-Based Management. *Environmental Impact Assessment Review, 16*, 71-102.
Dearden, P., and J. Dempsey. 2004. Protected Areas in Canada: A Decade of Change. *Canadian Geographer, 48*, 225-39.
Ducks Unlimited Canada. 2004a. Our Wetland and Wildlife Conservation Progress. http://www.ducks.ca/aboutduc/progress/index.html.
–. 2004b. Research Guiding Conservation. http://www.ducks.ca/aboutduc/how/research.html.
Francis, G. 2003. Governance for Conservation. In F.R. Westley and P.S. Miller (Eds.), *Experiments in Consilience: Integrating Social and Scientific Responses to Save Endangered Species* (pp. 223-379). Washington, DC: Island Press.
Halvorsen, K.E. 2001. Relationships between National Forest System Employee Diversity and Beliefs Regarding External Interest Groups. *Forest Science, 47*, 258-69.
Heynen, N., and P. Robbins. 2005. The Neoliberalization of Nature: Governance, Privatization, Enclosure and Valuation. *Capitalism Nature Socialism, 16*, 5-8.
Hurley P.T., and P.A. Walker. 2004. Whose Vision? Conspiracy Theory and Land-Use Planning in Nevada County, California. *Environment and Planning A, 36*, 1529-47.
Indigenous Governance Program. 2003. Protocols and Principles for Conducting Research in an Indigenous Context: Indigenous Governance Program, Faculty of Human and Social Development. Victoria, BC: University of Victoria. http://web.uvic.ca/igov/research/index.html.
Katz, C. 1998. Whose Nature, Whose Culture? Private Productions of Space and the "Preservation" of Nature. In B. Braun and N. Castree (Eds.), *Remaking Reality: Nature at the Millennium* (pp. 46-63). London and New York: Routledge.
Lane, M., 2003. Decentralization or Privatization of Environmental Governance? Forest Conflict and Bioregional Assessment in Australia. *Journal of Rural Studies, 19*, 283-94.
Lowe, P., J. Murdoch, T. Marsden, R. Munton, and A. Flynn. 1993. Regulating the New Rural Spaces: The Uneven Development of Land. *Journal of Rural Studies, 9*, 205-22.
Ludwig, D. 2001. The Era of Management Is Over. *Ecosystems, 4*, 758-64.
Magnusson, W., and K. Shaw (Eds.). 2003. *A Political Space: Reading the Global through Clayoquot Sound*. Minneapolis: University of Minnesota Press.
McGillivray, W.B. 1998. The Aspen Parkland: A Biological Perspective. In D.J. Goa and D. Ridley (Eds.), *Aspenland, 1998: Local Knowledge and Sense of Place* (pp. 95-103). Red Deer, AB: Central Alberta Museums Network.
McLain, R.J., and R.G. Lee. 1996. Adaptive Management: Promises and Pitfalls. *Environmental Management, 20*, 437-48.
Mitchell, B. 2002. *Resource and Environmental Management*. (2nd ed.) New York: Prentice Hall.
Nature Conservancy of Canada. 2003. *Prairie Conservation Planning Update: Highlighting Ecoregional and Site Conservation Planning Across the Prairies by the Nature Conservancy of Canada and its Partners*. Regina: Nature Conservancy of Canada.
–. 2001. *Conservation Science: A Science-Based Conservation Program*. http://www.natureconservancy.ca/files/frame.asp?lang=e_®ion=1&sec=science.
–. *n.d. A Conservation Toolchest for Landowners: Investing in the Land You Love*. http://www.natureconservancy.ca/pdf/tool%20chest%20revised.pdf.
Pulido, L. 1996. *Environmentalism and Economic Justice: Two Chicano Struggles in the Southwest*. Tucson: University of Arizona Press.
Reed, M.G. 2007. Uneven Environmental Management: A Canadian Perspective. *Environmental Management, 39*, 30-49.

–. 2003. *Taking Stands: Gender and the Sustainability of Rural Communities.* Vancouver: UBC Press.
–. 1997. The Provision of Environmental Goods and Services by Local Non-Governmental Organizations: An Illustration from the Squamish Forest District, Canada. *Journal of Rural Studies, 13,* 177-96.
Reed, M.G., and J. McIlveen. 2006. Toward a Pluralistic Civic Science? Assessing Community Forestry. *Society and Natural Resources, 19,* 591-607.
Roberts, R., and J. Emel. 1992. Uneven Development and the Tragedy of the Commons: Competing Images for Nature-Society Analysis. *Economic Geography, 68,* 249-71.
Robertson, M. 2005. The Neoliberalization of Ecosystem Services: Wetland Mitigation Banking and Problems in Environmental Governance. *Geoforum, 35,* 361-73.
Saskatchewan Watershed Authority. 2003. *Hydrographic Data for Redberry Lake near Krydor Levels 1966 – 2003,* Station 05GD003.
Schlosberg, D. 2004. Reconceiving Environmental Justice: Global Movements and Political Theories. *Environmental Politics, 13,* 517-40.
Schmutz, J. 1999. *Community Conservation Plan for the Redberry Lake Important Bird Area.* Saskatoon: Centre for Studies in Agriculture, Law and the Environment, University of Saskatchewan.
Sian, S. 2001. *Redberry Lake Biosphere Reserve: A Community's Plan for Sustainability.* Hafford, SK: Redberry Lake Biosphere Reserve. http://www.redberrylake.ca/images/Adobe%20documents/Redberry_Lake_Coop_Plan.pdf.
Smith, N. 1984. *Uneven Development: Nature, Capital and the Production of Space.* Oxford: Blackwell.
UNESCO, 2000. *Solving the Puzzle: The Ecosystem Approach and Biosphere Reserves.* Paris: United Nations Education, Scientific and Cultural Organization.
Ward, N., P. Lowe, S. Seymour, and J. Clark. 1995. Rural Restructuring and the Regulation of Farm Pollution. *Environment and Planning A, 27,* 1193-211.
Young, I.M. 1990. *Justice and the Politics of Difference.* Princeton, NJ: Princeton University Press.

10
Framing Environmental Inequity in Canada: A Content Analysis of Daily Print News Media

Leith Deacon and Jamie Baxter

Environmental inequity takes different forms, for example as procedural inequity or as outcome inequity. In this chapter we propose that environmental inequities extend into media coverage of pollution exposure. We suspect there is a substantial disconnect between the stories that are featured, or more often *not* featured, in the news media related to environmental inequity and the on-the-ground realities. For instance, there is a considerable amount of academic literature showing inequities – that blacks, Hispanics, and low-income groups are disproportionately exposed to pollution – in the United States (Bowen, 2002; Mohai and Bryant, 1992; Paulido, 1996; United Church of Christ, 1987), yet we expect that news media coverage of those same inequities are disproportionately low. Similarly, in Canada there are pollution issues that do receive the attention of academics (Haalboom, Elliott, Eyles, and Muggah, 2006; Lambert, Guyn, and Lane, 2006; Lambert and Lane, 2004), yet it is unclear if the national media are likewise interested in this issue. Although Canadian environmental inequity research is less developed than in the United States, studies such as Buzzelli, Su, Le, and Bache (2006), Jerrett et al. (2001), and Jerrett, Eyles, Cole, and Reader (1997) nevertheless point in the same direction – that substantial socio-economic inequities in particular do indeed exist here. These studies are well conceived, but their coverage is limited to Vancouver, Hamilton, and Ontario respectively. Such case studies are important, but we would like to retract the lens in order to gain a better understanding of patterns of localized environmental conflict over pollution hazards *across* Canada. We do this with reference to those issues that are reported in the national Canadian print news media. Before providing a short discussion of environmental reporting in the media, we elaborate the rationale for doing a content analysis of the media by briefly reviewing the environmental justice and equity literature.

In this chapter we use Draper and Mitchell's definition (2001) of environmental equity: "The equal treatment and protection for various racial, ethnic and income groups under environmental statutes, regulations, practices, and elements of fairness and rights that are applied in a manner that yields no substantial differential impacts relative to the dominant group – and the conditions so created" (p. 94).

Environmental equity is often further conceptually split into procedural equity and outcome equity. These reflect the rules and practices that support or mitigate against public participation in decisions that have a pollution component in the former instance, and the end point spatial and social distribution of pollution and people in the latter (Cutter, 1995). Procedural equity emphasizes the root causes of environmental inequities and is the less studied of the two forms of equity. Both may include temporal and spatial elements, but for the most part studies to date have been cross-sectional and/or case-study based (e.g., see Bowen, 2002).

Although most of the evidence supports the idea that inequities are a fact of life in North America, there are some contrary findings, most notably Anderton, Anderson, Oaks, and Fraser (1994), and serious methodological criticisms have been levelled against some of the earliest work in particular (e.g., see Bowen, 2002). We do not plan to wade into these debates; rather, we provide a methodological alternative (media content analysis), in a new context (Canada), that adds to the mix of research that has been dominated by GIS-supported regression analyses. Although we use similar techniques in other studies, this chapter seeks to first answer some basic questions about public dialogue on the juxtaposition of pollution and people in Canada. To be clear, we are not studying either procedural or outcome equity per se; we are studying the role being played by a stakeholder in the domain of environmental pollution justice. The nature of public media discussion on environmental inequity will provide an important window on *some* public discourses of pollution in Canada, and importantly, the silences that go along with them. One of the key implications of such an exploratory study is to highlight the extent to which the media have been a potential force in the rectification of inequities, or a potential part of the problem itself through unequal reporting.

The influence of the news media as a transformative force in society has been well documented (Curtin and Rhodenbaugh, 2001; Dispensa and Brulle, 2003; Hoffman and Ocasio, 2001; Kenix, 2005; Kensicki, 2004). The media have been portrayed pejoratively as both active claims makers trying to set public agendas and as more passive, but nevertheless powerful, gatekeepers of information. Even in an area of increased democratization of information through internet blogs and other websites, the traditional media remain the primary source of information for the public and have a significant influence on our perception and knowledge of topics ranging from health to technology to the environment. It is in this guise that the media, and those who are responsible for determining its content and coverage, potentially shape our perception of what is socially important and, in turn, what is considered news. This control can be abused by enabling powerful interest groups to transform the mass media into an arena that defines and frames social issues and concerns, arguably in a self-serving interest. Indeed, this abuse is a common claim against more conservative news agencies that align with large corporate interests (Chomsky and Herman, 1988). However, by the same token, the majority of media in North America have recently been accused of a generally liberal political bias (Goldberg, 2003) – an accusation that has sparked hot debate in the communications literature (Alterman, 2003) – and may potentially send conservative-leaning publications farther to the right of the political centre.

Research and policy concerning environmental inequity is well developed in the United States compared with Canada, but we have to be somewhat cautious about relying on the media as the *primary* barometer of environmental equity issues. The momentum for the environmental equity movement grew partially out of the struggle for civil rights in the United States, as far back as the 1960s. Further, a culmination of environmental disasters such as Love Canal, Three Mile Island, and Chernobyl resulted in widespread media attention on environmental contamination during the 1970s and 1980s. Although Canada has its own share of environmental disasters, such as the Sydney tar ponds, these do not appear to have the same international, or even national, profile as similar events from the United States and elsewhere. We are not sure why. Temporal trends in American-based media research has documented that the public's interest in, and media coverage of, environmentally related events climaxed after the 1989 Exxon Valdez oil spill (Mazur, 1998). However, the danger in relying solely on the media as a yardstick of the seriousness of environmental inequities is that other issues can supplant such coverage, at least for a time. For example, though environmental problems were displaced from the forefront of news coverage by the start of the Gulf War in 1991 and the collapse of the Soviet Union, this does not necessarily mean environmental inequity problems were being quietly rectified in the background. While the 1990s ushered a new "environmentally conscious" administration into the White House, American news coverage of environmental issues waned (Mazur, 1998). Even the formation of an Environmental Equity Work Group and President Clinton's executive order specifying environmental justice for minorities did not increase coverage at the time. We do not explore here why the coverage of certain issues rises and falls in the media, but we do want to highlight that all coverage must be set in a wider context of issues important to the media and society at large.

Although there is certainly a growing body of research on environment and the media generally (e.g., Clark, Barton, and Brown, 2002; Curtin and Rhodenbaugh, 2001; Shibley and Prosterman, 1998), little concerns environmental equity specifically. Further, American studies outnumber Canadian studies. Nevertheless, there are useful Canadian examples, such as Wakefield and Elliott (2003), which includes an examination of the role of local newspapers in environmental risk communication, and Driedger and Eyles (2003), concerning the Canadian media's framing of health risk from chlorination by-products in drinking water. While Wakefield and Elliot do refer to equity in the findings, their discussion does not concern the spatial or social distribution of threat. We could find no studies that explicitly examine Canada's news coverage of environmental inequity as the distribution of pollution relative to specific social groups, particularly marginalized peoples.

In this chapter we address this gap with an exploratory content analysis of Canada's two national daily newspapers, examining how environmental equity coverage varies over time and space and how it is framed in the Canadian context. This analysis is part of the first phase of a three-phase study meant to broaden and deepen our understanding of environmental equity in Canada. Although the first phase of the study is meant to identify inequity hotspots for detailed study in the later two phases, this content analysis plays largely a peripheral, supporting role regarding hotspots. That is, this analysis is not simply a tool for narrowing the focus to a few case-study communities for the larger study; rather, it achieves important objectives of its own, as follows:

(1) to identify the relative importance of environmental inequity coverage in the media
(2) to identify which types of environmental inequity coverage dominate
(3) to examine the temporal trends of environmental equity coverage
(4) to examine how coverage varies over space – by census metropolitan area and province
(5) to identify which stakeholder sources are cited most frequently and how this may bias coverage.

Methods

We use a content analysis of the two national daily newspapers, the *Globe and Mail* and the *National Post,* to address the five objectives. We use a traditional form of content analysis that involves the quantification of the *manifest* content of the newspaper articles. This may be contrasted with discourse analysis, which puts considerably more emphasis on the underlying meanings of media texts – their *latent* content (Krippendorff, 2003). Thus, we sacrifice depth for breadth, since this study is meant to provide a broad picture of this particular form of environmental coverage in Canada.

Content analysis is a well-established research method (Berelson, 1952; Fearing, 1953; Kerlinger, 1964) and can be defined as "a research technique used to make replicable and valid inferences from text (or other meaningful matter) to the contexts of their use" (Krippendorff, 2003, p. 18). Content analysis enables us to examine a large amount of textual information and identify some of its key properties – e.g., word frequency – to make inferences about the message itself, the audience, and/or the sender(s) of the message (Weber, 1990). In our case, the focus is mostly on the message and the senders, in the sense that we define the overall shape of coverage over time and space, including gaps in coverage.

Objectivity, systematization, and *reliability* are the strengths typically touted for quantitative content analysis. Objectivity provides scientific credence, specifying precisely defined categories that allow similar conclusions to be reached by different analysts following identical procedures with the same set of data. Systemization refers to the inclusion or exclusion of categories that follow a set of coherent principles, while reliability, which is closely connected to objectivity and systematization, refers directly to the repeatability of findings under similar procedures.

Nevertheless, it is challenging to locate environmental inequity coverage using keyword or subject searches because terms such as *equity* are more often used in the newspapers for issues of financial equity, rather than being social justice related. For this reason, we use a relatively broad definition of *in*equity coverage in this analysis. That is, we define environmental equity coverage as those articles that involve an *exposure* element and an *equity* element together (see Figure 10.1). The exposure element involves the words *environment* or *health* together with some word identifying pollution or contamination. The inequity component falls into four broad categories representing some of the key themes in the literature – general fairness, conflict, ethnicity/race, and socio-economic status.

A further complication is that, though a group may feel unjustly treated in terms of its actual or potential exposure to an environmental hazard, that does not necessarily mean an environmental injustice exists – even according to our rather broad definition. Although a community may complain about a hazard publicly, it may also be one of the main beneficiaries, in the form of, for example, stable local jobs. As another cautionary example, Baxter, Eyles, and Elliott (1999) report that relatively wealthy urban-fringe residents may feel that their community being the preferred site for a large municipal landfill is unjust. Yet the authors contrast this with the competing options that included both taking the waste to a much poorer community to the north and, the end result, continuing to truck local waste to relatively poor communities in the United States. Nevertheless, in order to err on the side of being inclusive, such examples if reported in the media would show up in this analysis. That is, in the article searches

Figure 10.1

Environmental injustice media analysis framework

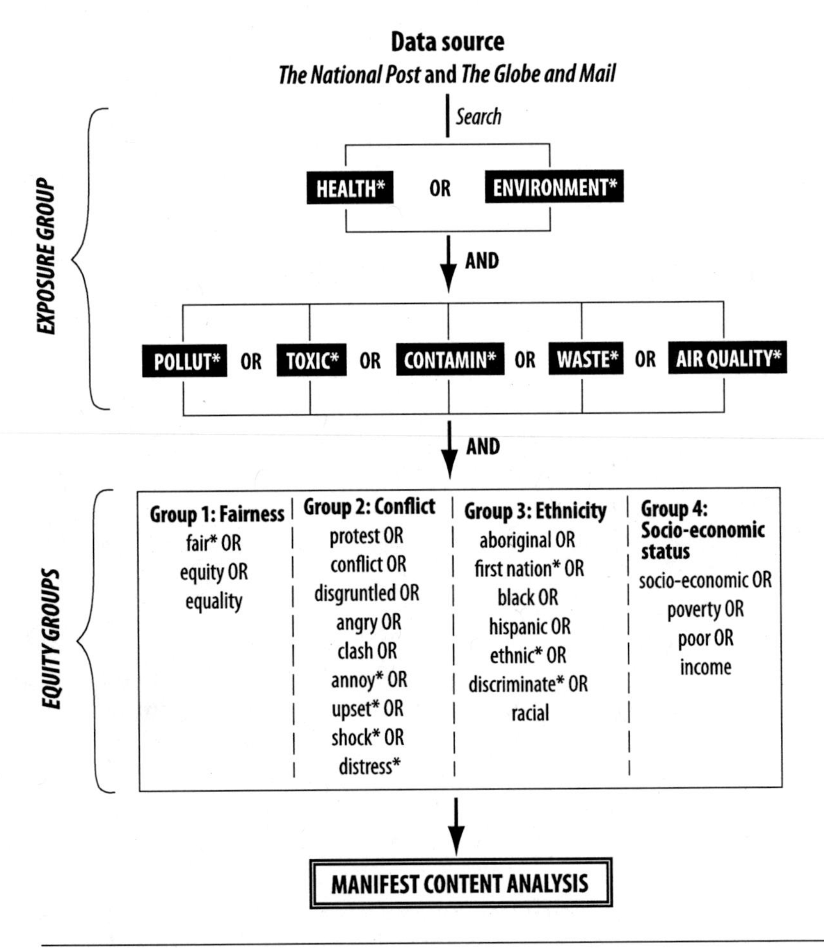

* indicates all synonyms were included in searches.

we do not distinguish between what may be termed *felt* inequity and some objective measure of what *we* feel is an injustice according to our definition or any other.

The data consist of news articles published in the *Globe and Mail* and the *National Post* between September 1986 and September 2006 and located using Factiva media database searches. The time period closely aligns with the rise of the environmental justice/inequity literature in academia. Keywords were searched only in the headline and lead paragraph to increase the likelihood that the focus of the article is environmental equity related.

Nevertheless, the unit of analysis is actually the full article. The *Post* and the *Globe* were selected because these newspapers compete for the same national audience and draw from the same newswire sources. They are the only two daily national print newspapers in Canada, and while both are regarded as conservative, the *Globe* is considered much less so than the *Post.* The *Globe and Mail,* founded in 1844, is Canada's oldest and largest national daily newspaper, with a weekly circulation of approximately 2,014,441. The *Globe* is generally regarded as socially liberal and fiscally conservative, with a reputation as being the voice of the central Canada elite, in particular Toronto's Bay Street financial community. The *National Post,* founded in 1998, is a much younger paper but has quickly reached a weekly circulation of 1,386,472, just over half of that of the *Globe.* The *Post* was established by Conrad Black as a conservative alternative to the traditional centre-left ideological stance found in most Canadian newspapers and often features columns by prominent neo-conservatives from both Canada and the United States. Thus, these papers together represent a reasonable cross-section of print media coverage in Canada.

Relevant articles were identified by the occurrence and co-occurrence of keywords in the headline or lead paragraph excluding republished news, recurring pricing, and market data, as well as other irrelevant sections such as the obituaries, sports, and calendars. As indicated above, each search contained the exposure component and one equity component. To narrow the possible range of stories to those involving pollution with a potential direct negative human outcome, the exposure group consists of *health** or *environment** and *pollut** or *toxic** or *contamin** or *waste** or *air quality**. Each search included wildcards "*" to account for variations of each term, for example, *fair** = *fair, fairness, unfair; contamin** = *contaminant, contamination, contaminate.* This second subset was initially just three terms, but the latter two were added after initial searches indicated that stories containing these terms frequently concern human health threats, even when the word *pollution, toxic,* or *contamination* is absent. Within each equity group, related terms were included in the search. The four equity groups, and their related words in italics, are: (1) fairness *(fair*, equity, equality),* (2) conflict *(protest, conflict, disgruntled, angry, clash, annoy*, upset*, shock*, distress*),* (3) ethnicity *(aboriginal, first nation*, black, hispanic, ethnic*, discriminate*, racial),* and (4) socio-economic status *(socio-economic, poverty, poor, income).* These terms relate to our initial definitions of environmental equity and equity coverage by emphasizing fairness, ethnicity, and socio-economic status. Conflict-related terms are also important to include, since the advancement of the environmental equity movement has historically been witness to elements of conflict (e.g., Bullard, 1990).

To create data for objective two, the identification of types of environmental inequity coverage dominate, a secondary block of searches was completed for Canadian census metropolitan areas (CMAs) with populations of a hundred thousand or greater, as well as all provinces and territories. CMAs, provinces, and territories were included in a search against the exposure group to determine the spatial location of pollution coverage across Canada.

The resulting database, which consists primarily of article frequencies, was used to test five categories of hypotheses, listed in Table 10.1. These are linked directly to the five objectives list above to explore how environmental equity coverage varies over time and space and how it is framed in the Canadian context. The hypotheses themselves, though supported by the literature, come also from our own experiences consuming Canadian news. Again, this is an exploratory study, and these hypotheses serve both a heuristic, organizational role and a deductive, theory-testing one.

The *prominence* hypothesis concerns our central assumption that there is very little coverage of environmental inequity in the media, especially in these largely centre-conservative daily newspapers. As it is difficult to set a cut-point for an expected number of articles, we compare the counts of our environmental inequity finds to those of common issues in the media – health care, un/employment, AIDS, and terrorism. These topics have been a staple in Canadian daily news for many years. To further detect if any lack of coverage is confined to a particular period, these contextual topics were searched along with environmental aggregate inequity – the exposure group *and* any of the inequity groups and their corresponding terms – for coverage over five-year periods starting in 1986. The *temporal* hypothesis is highly connected to the prominence hypothesis, but the former specifically tests whether coverage of environmental inequity increases over time starting in 1986 in five-year time brackets, with the count starting in September.

The *type of (environmental) inequity* coverage we expect will dominate is that concerning socio-economic status – the poor being disproportionately and unjustly exposed to pollution. That lower socio-economic groups tend to bear a disproportionate burden of pollution is a consistent finding in most of the environmental equity and justice literature (e.g., General Accounting Office, 1983; Mohai and Bryant, 1992; United Church of Christ, 1987). We expect the ethnicity block shown in Figure 10.1 to run a close second since predominantly black, and particularly Hispanic, neighbourhoods are often found to be statistically significant predictors of high pollution neighbourhoods (e.g., Mohai and Bryant, 1992; Zimmerman, 1993). Regardless, in the Canadian context ethnic and racial segregation is far less pronounced. As Figure 10.1 indicates, the only specific group we identify in our keyword searches in addition to black and Hispanic is First Nation/Aboriginal. The reasoning is twofold. First, the variety of possible groups is too large to include in a study such as this. Second, First Nations are highlighted because many are still very connected to their environ-

Table 10.1

Hypotheses and objectives

Theme	Hypothesis	Objective
Prominence	Coverage of environmental inequity will be scant when compared with coverage of other issues such as health care, un/employment, AIDS, or terrorism.	1
Temporal	Coverage of environmental inequity will increase over time.	2
Type of inequity	The most frequent coverage will be for the socio-economic form of environmental inequity.	3
Spatial	Stories in larger urban centres will dominate the coverage. Stories from the CMAs (population of 100,000+), particularly the most populated CMAs of Toronto, Montreal, and Vancouver, will dominate; stories from the five provinces with the highest populations (ON, PQ, BC, AB, MB) will dominate the coverage.	4
Cited sources	Government officials and industry representatives will be the most prominent sources.	5

ments vis-à-vis, for example, country foods and cultural practices – they have a unique sensitivity to environmental degradation.

Our assumption regarding the *spatial* distribution of coverage – that the large centres will dominate coverage – is that of somewhat of a contrast between the academic literature and popular media. The former includes famous and important case studies of rural inequities (e.g., Bullard, 1994), while the national print media tend to centre on big cities. Nevertheless, several empirical studies of environmental equity do include large centres (e.g., Mohai and Bryant, 1992). That is, we expect a tension between the need to report the plight of those fighting to prevent environmental harms and the need of the popular press to sell newspapers.

We expect a bias in the news media's reliance on certain *cited sources*, notably government and industry representatives. This hypothesis is supported by previous media analysis research (Sutter, 2001) that examines if particular sources (e.g., industry and government) are more frequently

Figure 10.2

Aggregated contextual media coverage in Canada, 1986-2005

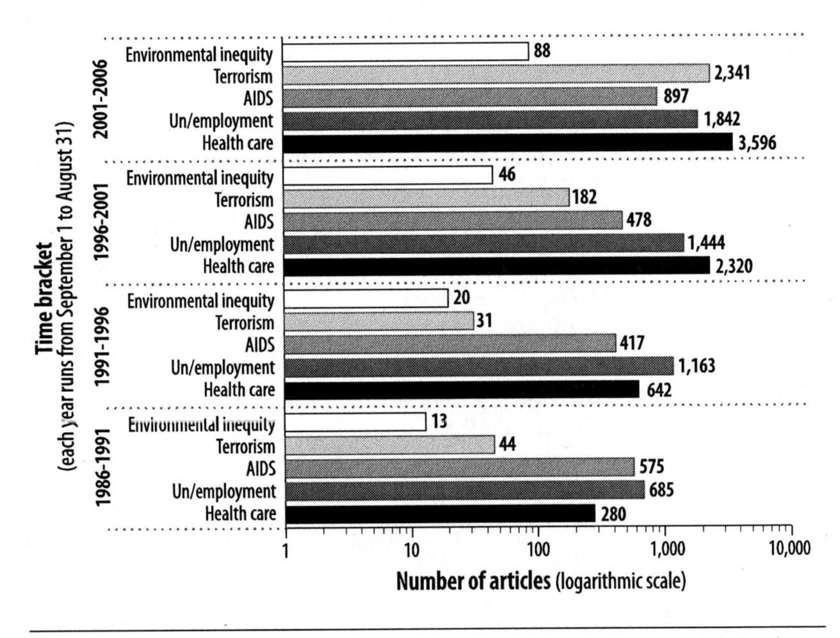

Note: The differences between time periods for each issue are significant at the 0.01 level.

cited than others (e.g., local stakeholders). This disproportional reliance favouring government and industry is often seen as more environmentally conservative, which would significantly impact how environmental inequity coverage is presented in the national print news media.

Results

Results are presented in order of the hypotheses outlined in Table 10.1.

Prominence

Figures 10.2 and 10.3 illustrate that environmental inequity is indeed the least frequently reported issue of the contextual-issue set we chose. What is important to note is that the scale presented is logarithmic, meaning the actual differences are very large. Environmental inequity coverage, with a total of 167 articles in the past twenty years, is 1,317 percent orders of magnitude lower than the least frequently reported issue from the rest of the group, AIDS, with 2,367 articles. Comparing the coverage of environmental inequity to watershed issues such as terrorism, unemployment, AIDS, and health care may seem an unfair comparison, however, these

Figure 10.3

Total contextual media coverage in Canada, 1986-2005

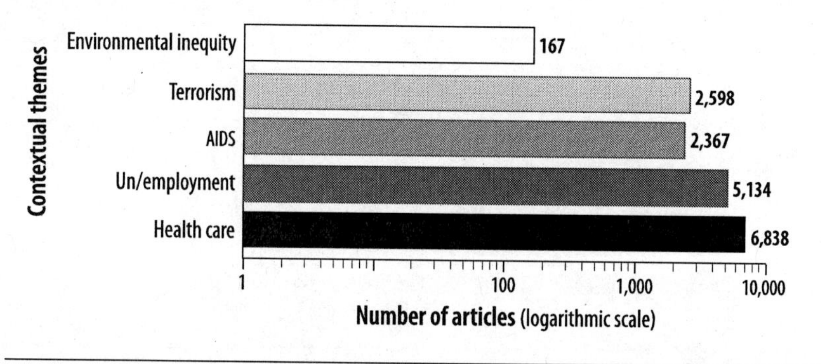

Note: The differences between contextual themes are significant at the 0.01 level.

issues are common themes in Canadian print news media and contrasting their coverage against environmental inequity coverage over the past twenty years adds context. Further, Figure 10.5, below, helps to compare more precisely how prominence has changed over time.

Temporal

We expected environmental inequity coverage to increase over time, and Figure 10.4 shows that this is indeed the case. The article count consistently increased: 13 articles in the 1986-1991 bracket, 20 articles in the 1991-1996 bracket, 46 articles in the 1996-2001 bracket, and 88 during the 2001-2006 bracket – which represents percentage increases of 54, 130, and 91 percent respectively. However, while coverage of environmental inequity issues has consistently increased and at a higher rate than many issues in any given five-year period, it appears that environmental inequity will always be outpaced by some other issue in the Canadian news. Figure 10.5 represents the aggregated percentage increase in the amount of coverage between five-year time blocks over the last twenty years, compared with four contextual themes. It reveals that environmental inequity coverage jumped 54 percent between the first and second time brackets (1986-1991 and 1991-1996), from 13 articles to 20 articles. There was then a significantly higher 130 percent increase in coverage between the second and third time bracket (1991-1996 and 1996-2001), here from 20 to 46 articles, followed by a significantly lower 91 percent increase in coverage between the third and fourth time bracket (1996-2001 and 2001-2006), here from 46 articles to 88 articles. This increase is comparable to the four contextual issues; however, it should be noted that terrorism-related cov-

Figure 10.4

Media coverage of aggregated environmental inequity in Canada, 1986-2005

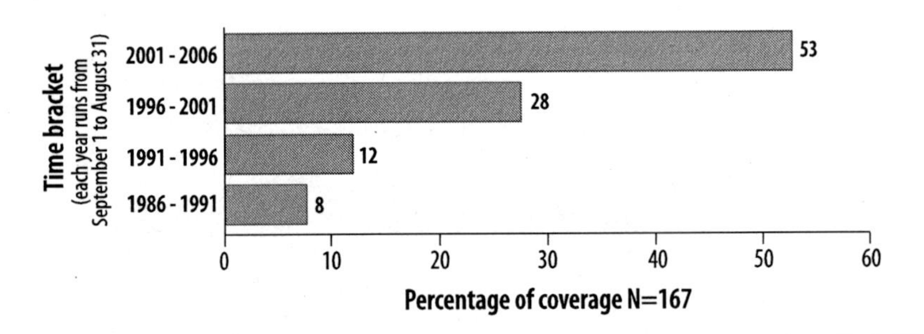

Note: These results are significant at the 0.05 level.

Figure 10.5

Aggregated percentage increase of contextual media coverage in Canada, 1986-2005

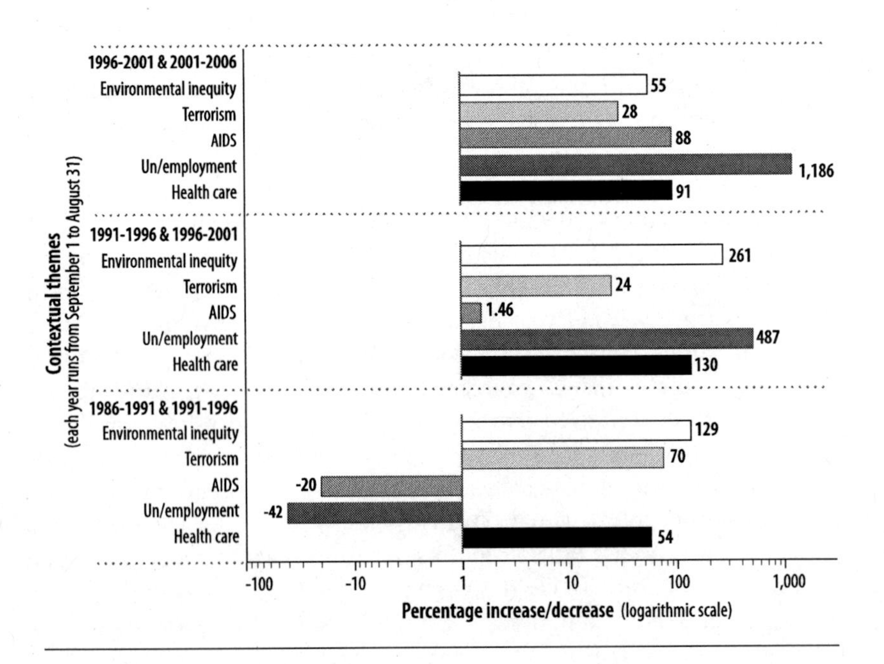

Figure 10.6

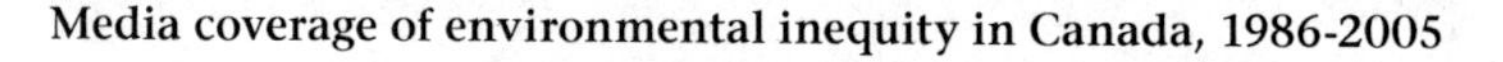

Media coverage of environmental inequity in Canada, 1986-2005

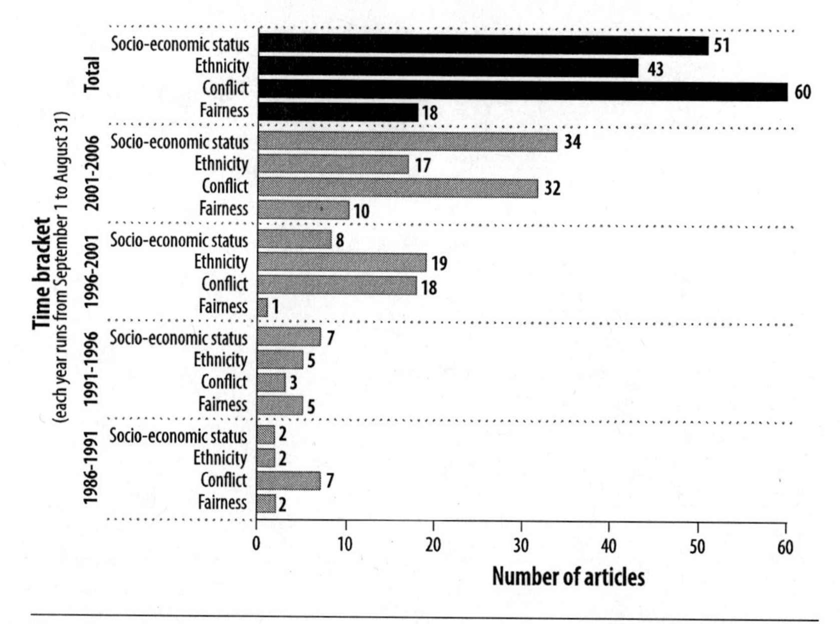

Note: The results between time periods for each issue are significant at the 0.01 level.

erage experienced the greatest single jump, over 1,100 percent, between 1996-2000 and 2001-2005. This can be assumed to be related to the events of 9/11. So, the increase in coverage for environmental inequity keeps pace with most of these marquee Canadian issues, albeit in relative terms only.

Type of Inequity

Our primary hypothesis regarding the types of inequity covered is that stories concerning socio-economic status environmental inequity – poorer people being disproportionately exposed to pollution – will dominate coverage. Although this hypothesis derives from findings in the academic literature, Figure 10.6 shows that this hypothesis seems not to be supported as stories related to conflict accounted for the greatest percentage of coverage (N=172) at 34 percent; followed closely by socio-economic status (30 percent), ethnicity (25 percent), and fairness (10 percent). Though the difference is statistically significant, stories on socio-economic status do run a very close second, followed next by ethnicity environmental inequity reporting. Interestingly, any direct mention of fairness and equity as represented by the fairness block runs a distant fourth overall.

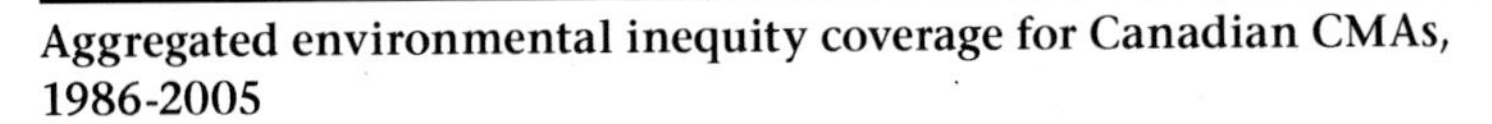

Figure 10.7

Aggregated environmental inequity coverage for Canadian CMAs, 1986-2005

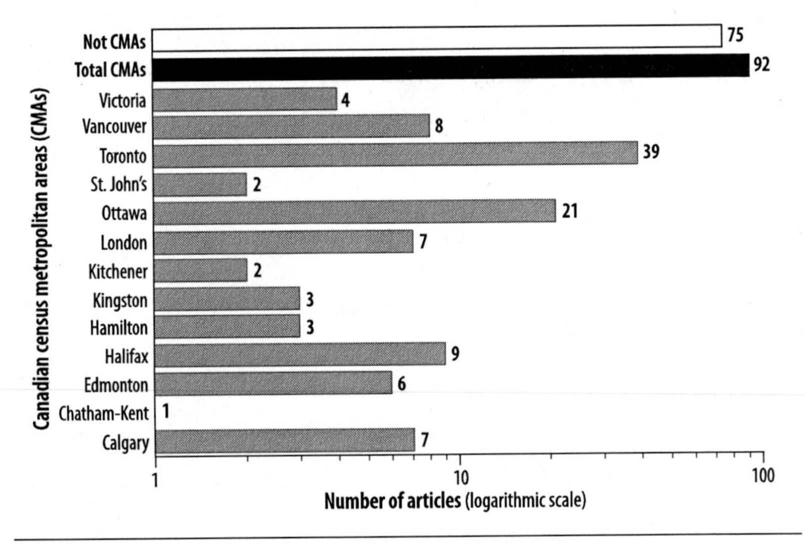

Note: The results among the CMAs are significant at the 0.01 level.

Spatial

In terms of spatial coverage of environmental inequity in the Canadian print news media, we expected major urban centres to dominate coverage. There are at least three indicators, shown in Figure 10.7, in support of this hypothesis. First, Toronto represents nearly 50 percent of all coverage in CMAs with populations over 100,000. This was distantly followed by Ottawa-Hull with 19 percent, and Vancouver with 13 percent.

Second, when environmental inequity articles from 2001-2005 that mention *any* CMA (over 100,000) are compared with those that do not mention any of these CMAs, the counts are comparable, at 48 articles and 42 articles respectively. These numbers may be proportional to the populations of CMAs versus non-CMAs, but this is disproportional to the number of cities in Canada. For example, in the 2001 census there were 46 cities in Canada over 50,000 people (census agglomeration areas or CAs), 19 of which are not CMAs over 100,000, and 27 of which are CMAs. Since there were inequity media articles on only 13 of the CMAs, this represents ratios of 48 articles to 13 large cities versus 42 articles to 43 smaller cities. This latter figure is, of course, the most optimistic estimate since the non-CMA articles may not even refer to a CA at all but, rather, to any smaller town. Clearly, the preponderance of reporting is in these large centres.

Figure 10.8

Aggregated cited sources of environmental inequity-related articles in the Canadian daily media, 2001-2005

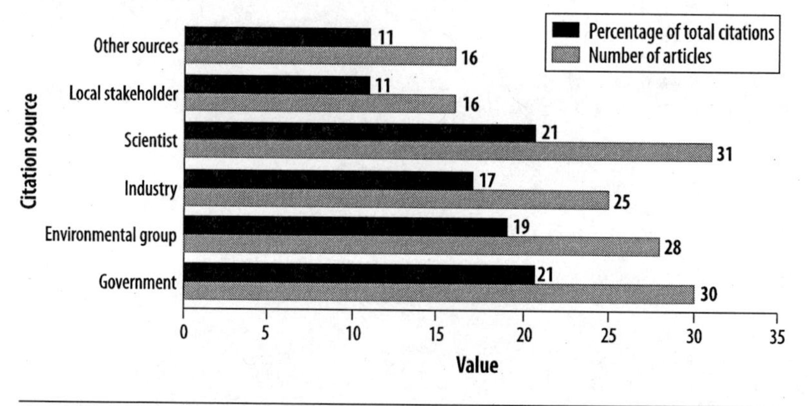

Third, of the thirteen CMAs with populations over 100,000 that were featured in the daily news media, CMAs located in Ontario represented 54 percent of coverage. While Ontario is Canada's most populous province with approximately 12 million people, this number is no where near 54 percent of the population of Canada. This indicates that there is a very high Ontario-bias of coverage regarding environmental inequity issues in Canada's daily print news media.

Cited Sources

Our hypothesis related to source citation – the people quoted in an article – was that certain information sources would be relied upon disproportionately in the news media, and specifically that industry and government sources would be the most heavily relied upon. The corollary to this is that local stakeholders would have the smallest voice – representing yet another form of inequity in coverage. To be able to compare coverage between the *National Post* and the *Globe and Mail,* only newspaper articles from 2001-2005 were examined and vetted in this section, since this was the only time frame in which both papers were published.

As Figure 10.8 shows, we did indeed find that government accounts for the greatest percentage of citations (21 percent), but this is tied with the figure for scientists. Both of these groups were very closely followed by environmental groups (19 percent) and industry representatives (17 percent). The major standouts are on the low end, though – the local stakeholders and "other sources" – each of which account for only 11 percent of the quotations in these newspapers. When source citation was further broken down

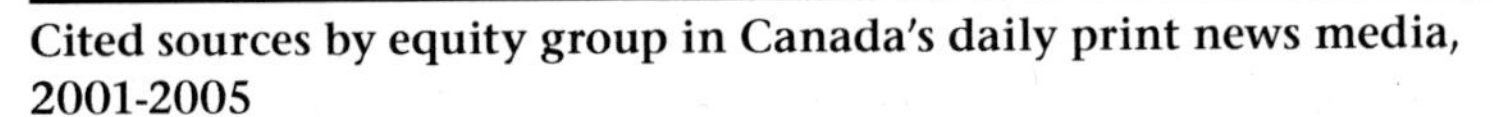

Figure 10.9

Cited sources by equity group in Canada's daily print news media, 2001-2005

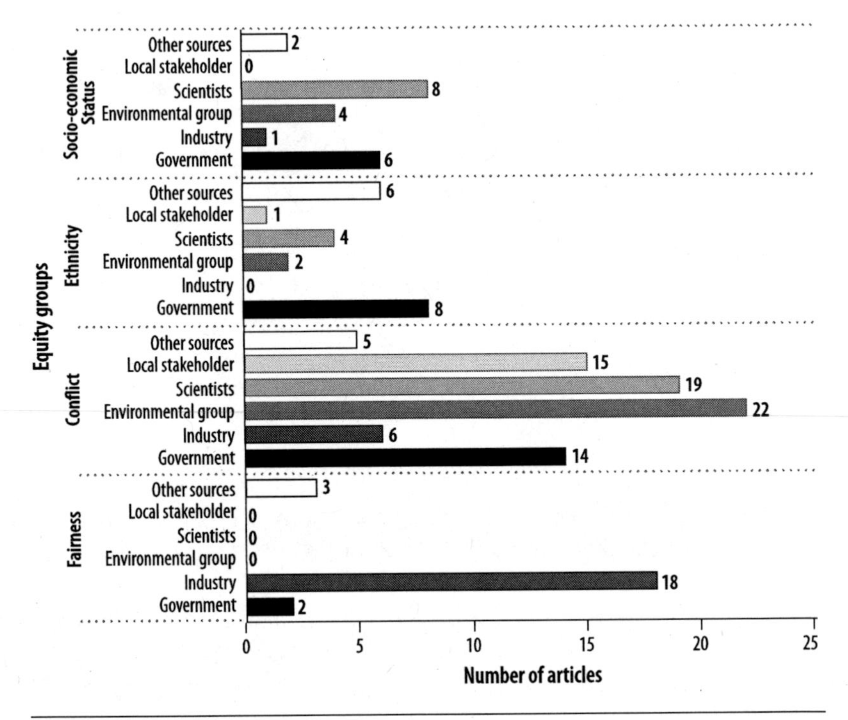

Note: These results are significant at the 0.05 level.

by equity group, we see that different sources dominate reporting on certain forms of inequity (Figure 10.9). We found that in the conflict equity group, environmental groups (27 percent) and scientists (23 percent), followed by local stakeholders (18 percent), are the most cited. Within the "fairness" group, the fact that industry representatives accounted for 78 percent of the citations is far more curious. In the ethnicity equity group, government representatives were the most cited, with 38 percent, and in the socio-economic status group, scientists were most often cited, with 38 percent also.

Discussion

This chapter presents evidence to support our proposition that environmental inequity is not simply restricted to procedural and outcome equity but, rather, that environmental inequities extend into media coverage of pollution exposure. We explore five primary hypotheses related to the distribution of environmental inequity coverage across time, space, and source, as outlined in Table 10.2.

Table 10.2

Results of hypotheses and sub-hypotheses

Theme	Hypothesis	Supported?
Prominence	Coverage of environmental inequity will be scant when compared with coverage of other issues such as health care, AIDS, or terrorism.	Yes
Temporal	Coverage of environmental inequity will increase over time.	Yes
Type of inequity	The most frequent coverage will be for the socio-economic form of environmental inequity.	No
Spatial	Stories in larger urban centres will dominate the coverage.	Yes
Cited sources	Government officials and industry representatives will be the most prominent sources.	No

Prominence

We found that the amount of coverage related to a set of contextual issues was significantly higher than the amount of coverage related to environmental inequity in Canada. This finding supported our primary prominence hypothesis that coverage of environmental inequity would be scant compared to a set of watershed issues. What is of interest here is that while coverage of inequity has increased over a twenty-year period overall, the increase seems to be diminishing when compared with previous periods. This decreasing coverage does not support previous findings that indicate newspapers have a general liberal political bias (e.g., Goldberg, 2003). Rather, there seem to be other issues dominating coverage no matter what – since in no period did environmental inequity – arguably a "liberal" issue – even approach the top most-written-about issues. Given the well-documented role of newspapers as a source of information (Curtin and Rhodenbaugh, 2001; Dispensa and Brulle, 2003; Hoffman and Ocasio, 2001; Kenix, 2005; Kensicki, 2004), the scarce coverage dedicated to environmental inequity in the Canadian newspapers is cause for concern.

Temporal

The findings from the prominence objective are directly related to the temporal hypothesis – that coverage of environmental inequity will increase over time. As mentioned, there was an increase in coverage of environmental inequity-related stories over the past twenty years. This demonstrates that environmental inequity is becoming more recognized as a Canadian issue and an issue deserving of media attention and coverage. This is in contrast to American findings, which have noted that interest in environmental issues is waning (Mazur, 1998). What is important to note is that this increase is not uniform and appears to be in flux.

Type of Inequity

Our hypothesis that the socio-economic status form of environmental inequity would dominate coverage was not supported. This indicates a potential disconnect between the media and academia (e.g., Jerrett et al., 1997; Mohai and Bryant, 1992). A possible explanation for the finding that conflict inequity coverage dominates all others could be linked to both the number and the choices of keywords that relate to "conflict." Our keywords were designed to represent conflict themes that had been found in inequity literature. Yet given that there are nine keywords representing "conflict" and only three, seven, and four for fairness, ethnicity, and socio-economic status respectively, it is possible this has over inflated the counts for the conflict group. Nevertheless, since socio-economic status and pollution articles were the second most frequently reported form of inequity coverage, this disconnect may not be severe. Indeed, it may be that the cases involving conflict concern poorer communities railing against big business, for example. Testing such ideas, though, is well beyond the scope of this study, since the economic status of the parties in conflict is often not reported – and given the nonspecificity of most terms related to socio-economic status – e.g., *money* – it is unlikely that new keywords relating to socio-economic status would find these articles.

Spatial

Our hypothesis that coverage of environmental inequity would be focused in the major urban areas across Canada was verified. Toronto represents nearly 50 percent of all coverage and if Ottawa-Hull and Vancouver related coverage is included, this number increases to nearly 80 percent of all inequity coverage in CMAs with populations over 100,000. While the preponderance of reporting is in these largest centres, polluting developments are spread all over the country (NPRI, 2006). This discrepancy supports our theory that a tension exists between the need to report on the plight of those fighting to prevent environmental harms and the need of the popular press to sell newspapers. Further, there are important Canadian pollution events with associated equity dimensions that are seen as legitimate and important problems in the academic literature that may not be receiving proportional coverage in the media. For example, numerous articles have been written about the toxic Sydney tar ponds in Nova Scotia (Haalboom et al., 2006; Lambert et al., 2006; Lambert and Lane, 2004), yet this does not seem to be reflected in the national media coverage of this issue.

Cited Sources

Our main reason for including information about cited sources was to determine if local stakeholders have the same voice as others in the media. It appears that they do not. The reliance on government and scientists as sources of information, both of which are often viewed as environmentally conservative, would significantly impact how environmental inequity issues are framed in Canada and possibly give support to nonenvironmentally friendly practices. Within each of the equity groups reliance upon one or another information source varied and no one particular group dominated – with the exception of the fairness equity group. Industry sources accounted for nearly 90 percent of citations for this group – something very surprising to us. This finding may indicate that government sources are perceived as unbiased on the other categories and possibly fairer in the eyes of the media.

It also is worth noting that environmental coverage varied according to national newspaper. Environmental articles were initially selected from each of the two papers if they contained inequity-related terms; however, as mentioned, such terms, which we have used to determine "inequity," may be conflated with other issues (e.g., equity associated with finances). This unvetted search resulted in the *National Post* accounting for 60 percent of inequity coverage (N=98). But once each article was vetted to ensure it was related to environmental inequity, the *Globe and Mail* accounted for 60 percent of this more precise article set (N=37).

Conclusion

This chapter demonstrates that environmental inequity extends into media coverage of pollution exposure in Canada. We use a method – content analysis – that is very different from what is typically used to study environmental inequity. Although many studies have examined equity in terms of procedural and outcome equity (e.g., Cutter, 1995), we reinforce what Haluza-DeLay, O'Riley, Cole, and Agyeman (this volume) call "recognition" by showing that key media give particular groups standing by way of disproportionate coverage. Media coverage may affect recognition, or more importantly a lack thereof, in several ways: (1) it may give prominence to certain issues while neglecting others; (2) because of increasing coverage of a certain issue over time, readers may come to believe that that particular issue is most important; (3) since the media appears to have a particular bias in favour of large metropolitan areas, smaller communities and *their* environmental inequity issues may be neglected; and (4) a newspaper's political bias may affect coverage by giving preference to certain issues and certain information sources (e.g., government, industry, scientific experts), while neglecting others (e.g., local stakeholders, members of the community).

Future research could focus not only on the manifest content of such media but also on the latent content. Further, a discourse analysis of interviews and focus groups of relevant stakeholders in any given environmental dispute would enable a better understanding of the underlying meanings of environmental threats. National coverage of these issues is important if common threads between inequity events are to be brought together – for example, to bring stakeholders regarding issue A in place B together with stakeholders regarding issue C in place D. Leaving it to the local papers would likely bypass this; it's a niche that national papers would best fill.

There is value in examining how environmental inequity is located in Canada's daily print news. We provide a methodological alternative using the Canadian context; which contrasts the majority of empirical studies, which focus on particular forms of distributive equity in the United States. This approach, however, is limited because it tells us little about how these issues are framed in the local context and the precise ways the media may legitimate some groups and not others. Although we demonstrate that particular populations and locations are disproportionately reported on, creating a potential disconnect between the stories featured in the Canadian national daily print news and the on-the-ground realities, it is the nature of those realities that deserve urgent and ongoing attention by environmental justice researchers.

References

Alterman, E. 2003. *What Liberal Media? The Truth about Bias and the News*. New York: Basic Books.

Anderton, D.L., A.B. Anderson, J.M. Oaks, M.R. Fraser. 1994. The Demographics of Dumping, *Demography,* 31 May, *2*(2), 229-48.

Baxter, J., J.D. Eyles, and S.J. Elliott. 1999. From Siting Principles to Siting Practices: A Case Study of Discord among Trust, Equity and Community Participation. *Journal of Environmental Planning and Management, 42*(4), 501-25.

Berelson, B. 1952. *Content Analysis in Communication Research*. Glencoe, IL: Free Press.

Bowen, W. 2002. An Analytical Review of Environmental Justice Research: What Do We Really Know? *Environmental Management, 29*(1), 3-15.

Bullard, R.D. 1994. *Dumping in Dixie: Race, Class and Environmental Quality.* 2nd ed. Boulder: Westview Press.

–. 1990. *Dumping in Dixie: Race, Class and Environmental Quality.* Boulder: Westview Press.

Buzzelli, M., J. Su, N. Le, T. Bache. 2006. Health Hazards and Socio-Economic Status: A Neighbourhood Cohort Approach, Vancouver, 1976-2001. *Canadian Geographer, 50*(3), 376-92.

Chomsky, N., and E.S. Herman. 1988. *Manufacturing Consent: The Political Economy of the Mass Media*. New York: Pantheon Books.

Clark, L., J. Barton, and N. Brown. 2002. Assessment of Community Contamination: A Critical Approach. *Public Health Nursing, 19*(5), 354-65.

Curtin, P.A., and E. Rhodenbaugh. 2001. Building the News Media Agenda on the Environment: A Comparison of Public Relations and Journalistic Sources. *Public Relations Review, 27,* 179-95.

Cutter, S. 1995. Race, Class, and Environmental Justice. *Progress in Human Geography, 19*(1), 111-22.

Dispensa, J.M., and R.L. Brulle. 2003. Media's Social Construction of Environmental Issues: Focus on Global Warming: A Comparative Study. *International Journal of Sociology and Social Policy, 2*(10), 74-105.

Draper, D., and B. Mitchell. 2001. Environmental Justice Considerations in Canada. *Canadian Geographer, 45*(1), 93-98.

Driedger, S.M., and J. Eyles. 2003. Difference Frames, Different Fears: Communicating about Chlorinated Drinking Water and Cancer in the Canadian Media. *Social Science and Medicine, 56,* 1279-93.

Fearing, F. 1953. Towards a Psychological Theory of Human Communication. *Journal of Personality, 22*(1), 71-88.

General Accounting Office (GAO). 1983. Siting of Hazardous Waste Landfills and Their Correlation with Racial and Economic Status of Surrounding Communities, *GAO/RCED-83-168.* Washington, DC: General Accounting Office.

Goldberg, B. 2003. *A CBS Insider Exposes How the Media Distorts the News.* New York: HarperCollins.

Haalboom, B., S. Elliott, J. Eyles, and H. Muggah. 2006. The Risk Society at Work in the Sydney "Tar Ponds" Risk. *Canadian Geographer, 50*(2), 227-41.

Hoffman, A.J., and W. Ocasio. 2001. Not All Events Are Attended Equally: Toward a Middle-Range Theory of Industry Attention to External Events. *Organizational Science, 12*(4), 414-34.

Jerrett, M., R.T. Burnett, P. Kanaroglou, J. Eyles, N. Finkelstein, C. Giovis, and J. Brook. 2001. A GIS-Environmental Justice Analysis of Particulate Air Pollution in Hamilton, Canada. *Environment and Planning A, 33,* 955-73.

Jerrett, M., J. Eyles, D. Cole, and S. Reader. 1997. Environmental Equity in Canada: An Empirical Investigation into the Distribution of Pollution in Ontario. *Environmental and Planning A, 29,* 1777-800.

Kenix, J.L. 2005. A Comparison of Environmental Pollution Coverage in the Mainstream, African, American and Other Alternative Press. *Howard Journal of Communications, 16,* 49-70.

Kensicki, L.J. 2004. No Cure for What Ails Us: The Media-Constructed Disconnect Between Societal Problems and Possible Solutions. *Journalism and Mass Communications Quarterly, 81*(1), 53-73.

Kerlinger, F.N. 1964. *Foundations of Behavioral Research: Educational and Psychological Inquiry.* New York: John Wiley.

Krippendorff, K. 2003. *Content Analysis: An Introduction to Its Methodology.* New York: Sage Publications.

Lambert, T.W., L. Guyn, and S.E. Lane. 2006. Development of Local Knowledge of Environmental Contamination in Sydney, Nova Scotia: Environmental Health Practice from an Environmental Justice Perspective. *Science of the Total Environment, 368,* 471-84.

Lambert, T.W., and S. Lane. 2004. Lead, Arsenic, and Polycyclic Aromatic Hydrocarbons in Soil and House Dust in the Communities Surrounding the Sydney, Nova Scotia, Tar Ponds. *Environmental Health Perspectives, 112*(1), 35-42.

Mazur, A. 1998. Global Environmental Change in the News 1987-90 vs. 1992-6. *International Sociology, 13*(4), 457-72.

Mohai, P., and B. Bryant. 1992. Environmental Racism: Reviewing the Evidence. In B. Bryant and P. Mohai (Eds.), *Race and the Incidence of Environmental Hazards, A Time for Discourse* (pp. 163-76). Boulder, CO: Westview Press.

Paulido, L. 1996. A Critical Review of the Methodology of Environmental Racism Research. *Antipode, 28*(2), 142-59.
Shibley, M.A., and A. Prosterman. 1998. Silent Epidemic, Environmental Justice, or Exaggerated Concern? *Organization – Environment, 11*(1), 33-58.
Sutter, D. 2001. Can the Media Be so Liberal? The Economics of Media Bias. *The Cato Journal, 20*(3), 431-51.
United Church of Christ. 1987. *Toxic Wastes and Race: A National Report on the Racial and Socioeconomic Characteristics of Communities with Hazardous Wastes Sites*. New York: United Church of Christ Commission for Racial Justice.
Wakefield, S.E.L., and S.J. Elliott. 2003. Constructing the News: The Role of Local Newspapers in Environmental Risk Communication. *Professional Geographer, 55*(2), 216-26.
Weber, R.P. 1990. *Basic Content Analysis*. Newbury Park, CA: Sage Publications.
Zimmerman, R. 1993. Social Equity and Environmental Risk. *Risk Analysis, 13*(6), 649-66.

11

Environmental Justice as a Politics in Place: An Analysis of Five Canadian Environmental Groups' Approaches to Agro-Food Issues

Lorelei L. Hanson

In *Forcing the Spring,* American environmental scholar Robert Gottlieb (2005) asserts that "food issues such as access, nutrition, price, and the overall system of production, distribution, sale and consumption of food products [have been] generally absent from the agendas of both the national environmental groups and local environmental justice groups" (p. 17). The popularity of books such as *Fast Food Nation* (Schlosser, 2001), movies such as *Super Size Me* (Spurlock, 2004), online videos such as *The Meatrix* (Sustainable Table, 2003), and demands for more robust food regulations and surveillance demonstrate considerable public interest and concern in the agro-food system. Yet environmentalists have been slow to take advantage of this opportunity to more fully explore the social and ecological dimensions of food issues by limiting their focus to examining single-issue topics, such as the toxicity of pesticides, herbicides, and fertilizers, or biotechnology. Clearly though, this approach is insufficient. As academic and popular works critiquing fast food, techno-foods, and global food-system trends by writers including Eric Schlosser (2001), Michael Pollan (2006), and Marion Nestle (2002) reveal, our apprehension about what we eat should not be limited to a specific food or agricultural practice. Attention needs to be directed at the food system as a whole as a crucial environmental and social justice issue (Gottlieb, 2005).

My objective in this chapter is to consider the accuracy of Gottlieb's portrayal of a lack of observance and critical analysis on the part of environmentalists of the environmental justice dimensions of the agro-food system within the Canadian context. I begin by engaging critically with the concept of environmental justice and, in particular, look at some of the applications of environmental justice discourse and frames of analysis that break away from the confines of its origins in US civil rights traditions. I then turn my attention to critically examining the agro-food strategies of five broad-based national environmental organizations in Canada: Greenpeace Canada, Friends of the Earth Canada, the David Suzuki Foundation, the Sierra Club of Canada, and the Canadian Environmental Network.[1] My intention is not to applaud or discredit the work of any one of these organizations. Rather, my interest is in examining the range of analysis, discourse, and approaches being adopted by environmentalists in addressing the complex social justice and environmental sustainability dimensions of agriculture and food issues in Canada. More substantially, my aim is to consider the limits and possibilities that exist in framing agro-food issues in Canada as matters of environmental justice.

Constructions of Environmental Justice beyond Race, Class, and the United States

As environmental justice activism and research has matured it has become apparent that the movement and analysis needs to become broader in scope and more encompassing in the sites, forms, and processes of injustice than it has been concerned with traditionally (Walker and Bulkeley, 2006). As Brulle and Pellow (2005) argue, "Scholars cannot understand – and policymakers prevent – environmental injustices through a singularly focused framework that emphasizes one form of inequality to the exclusion of others" (p. 297). Consequentially, the question "Justice for whom?" has become cast in more inclusive terms within environmental justice to include not only race and class but also differences of gender, ability, the rights of future generations and other species, and so on.

Critiques of the limited focus of environmental justice have been concerned not only with expanding who counts as a victim of pollution but also in developing a more relational and contextual frame of analysis regarding the context of the injustice. For example, Agyeman's notion (2005) of just sustainability is that environmental justice movements should take form according to their specific conditions and aspirations and as a part of this attention should be given to global-local linkages. Similarly, Teelucksingh (2002) argues that environmental justice scholarship and activism must consider "both structural and political economic processes and the manner in which various stakeholders act as agents in the reproduction of space reflecting their particular interests" (p. 122). The focus of most environmental justice activism continues to be at the local level, yet there is a need to extend the analysis of the circumstances of injustice beyond a site-specific, fixed notion of a "case in point" to a more relational view of space and social relations. As Debbané and Keil (2004) explain, this means that while we recognize environmental justice as a practice and concept as being locally grounded, we also need to embed the term in an understanding of the multiscalar "relationships of contradictions that create a unique event of environmental injustice at a given location" (p. 210). Geographies of political opportunities involving the juxtaposition of varied politics – local, regional, national, and global – shape environmental injustices and responses to them. Hence, the conceptual shift required is from an emphasis on environmental justice involving a politics of place to an emphasis on it emerging through a politics in place: "A distinctive politics of place based on the powers of proximity/particularity in a world of displaced and multiscalar happenings and power geometrics" (Amin, 2002, p. 397). Participants in the communities of concern, bearing multiple political identities, engage in these plural politics and multiple spatialities "in a narrative and material process of co-production of the world in which they live" (Debbané and Keil, 2004, p. 211).

In trying to account for the multiscalar and social complexity of these struggles, many environmental scholars have considered how the underlying causes and dynamics of these inequities manifest across temporal contexts, and cultural and national settings. Debbané and Keil (2004) contend that "Historical, social, cultural and political differences have created different arrays of state policies, civic organization, ideological hegemonies and legal realities in which the notion of environmental justice is embedded in each national context" (p. 212). For example, in contrast to the United States, where environmental justice is often engaged as a matter of civil rights, in Canada these issues are often framed through defence of the welfare state. Debbané and Keil assert that in Canada environmental justice "claims have less to do with individual or community rights but are linked, in large part, to the fear of the large segments of the population that the system of justice represented by public service delivery across this vast land may be threatened" (p. 218). Hence, you find Canadian scholars such Draper and Mitchell (2001) advocating that environment justice is best realized through governments in consultation and negotiation with local communities. Similarly, in her work on food security, Wekerle (2004) describes how transformative politics emerged within the food-security movement in Toronto, when it refocused their efforts from emergency food networks to the right to food as a component of a more democratic and just society. Key in addressing these injustices is the explicit recognition of the need to insist on community participation and the safeguarding of local control, but nonetheless there is an implicit trust and responsibility placed in governments to ensure that public goods such as environmental justice are provided and protected.

Draper and Mitchell's position may be naive and even problematic in its faith in a deliberative democracy that will manage the environment in an effective and nondiscriminatory way, but their and other Canadian scholars' comments point to several significant matters related to environmental justice in this country. They highlight the national variability within which matters of sustainability and justice are forged and contested, as well as stretch the notion of environmental justice to account for the state's responsibility for basic needs provisioning and environmental regulation. Practices of citizenship and state regulation as they relate to the public commons are thus highlighted by drawing attention to the material and structural practices and policies that divide groups by enabling or limiting their access to various resources, the discursive practices and ideologies that serve to justify these group hierarchies, and the level of citizen engagement required in creating transformative politics.

A Food-System Analysis of Five Canadian Environmental Groups

Canada has its share of environmental injustices within the agro-food sector. The hunger of hundreds of thousands of people (CAFB, 2006),[2] the exposure of farm workers to pesticides and other substances that have deleterious health impacts (CBC News, 2006; Choinière and Munroe, 1997), and the disproportionate risks and financial, social, and economic losses faced by farmers and ranchers throughout Canada (Hanson, 2007) are evidence of some of these. Simply put, matters of equity and sustainability are central issues to a long-term functional agro-food system.

To speak about the agro-food system is to draw attention to all the processes involved in feeding people: growing, harvesting, processing, packaging, transporting, marketing, distributing, consuming, and disposing of food (Wilkins and Eames-Sheavly, 2003). At each stage of this system there are myriad energy, chemical, human, financial, and other inputs and outputs, and all stages of the food system depend to some degree on networks of labour, research, and education. These networks take diverse forms, from the multimillion-dollar research conducted by Monsanto involving world-renowned scientists using field test plots of canola across Alberta farms, to the familial transmission of traditional ecological knowledge about indigenous foods and medicinal plants in the Hudson's Bay bioregion by Inuit and Cree (McDonald, Arragutainaq, and Novalinga, 1997). Each level of the food system thus operates within and is influenced to varying degrees by the social, political, economic, and natural environments within which it is situated and to which it is linked.

If greater social and ecological sustainability is to be achieved in this multifaceted and variable system, and opposition established that can effectively challenge and transform increased corporate-controlled global integration of it, links need to be forged between numerous food and agriculture advocacy groups (farmer groups, environmental organizations, development agencies, etc.). As a part of such an alliance, an integrated framework must be established that can collectively address the participant groups' distinctive concerns. As Buttel (1997) explains, it is only through broad coalitions composed of people with grievances about or interests in the current agro-food system that a sustainable agricultural movement will gain enough political and social force to "achieve the extent of meaningful impacts that are required to address fundamental agro-ecological sustainability problems" (p. 353). Because environmental justice concerns itself with the issues of daily life and has a strong tradition of community empowerment (Gottlieb and Fisher, 1996), it provides a conceptual discourse and set of strategic tools that are well suited to addressing issues of food security and sustainable agriculture. A relative and contingent form of environmental justice provides a valuable framework within which an expanded, multidimensional analysis of the space of food production, processing, distribution, and consumption could occur. For the purposes of this chapter, the question is, to what degree are Canadian national environmental groups engaging in this kind of activism and analysis?

Greenpeace Canada, Friends of the Earth Canada, the David Suzuki Foundation, the Sierra Club of Canada, and the Canadian Environmental Network are five of the most active and well-known Canadian national environmental groups that focus on a broad array of environmental issues. All these environmental nongovernmental organizations (ENGOs) undertake research, education, cooperation, and consultation, but they vary in their strategies. For example, Greenpeace members regularly engage in acts of civil disobedience, whereas the Canadian Environmental Network does not undertake direct advocacy activities; the remaining three ENGOs fall somewhere between these two poles. As well, the composition of the membership differs somewhat: unlike the other three ENGOs, the membership of the Canadian Environmental Network and Friends of the Earth Canada is composed of Canadian environmental groups rather than individual members.

I examined each of these five ENGOs' websites for materials that relate to food, agriculture, and environmental justice. Although the web postings do not offer a full view of the activities of these ENGOs, as one of the most public faces of these organizations, they provide an overview of their central activities and the image they are trying to project as they endeavour to increase public awareness, influence public opinion and government policy, and raise funds. In particular I was interested in their discursive constructions of environmental justice, and food and agricultural, and I read and critically analyzed any material that related to the social justice and environmental implications of the agro-food system. A summary of my findings is as follows.

Greenpeace Canada

- Has four official environmental campaigns, one of which is Say No to Genetic Engineering (GE) (Greenpeace, 2006).
- In its overview of GE, highlights the environmental and health risks of GE foods and advocates the labelling and segregation of GE seeds and crops.

- Opposes the patenting of plants, animals, humans, and genes, arguing that life is not a commodity.
- Compiled *How to Avoid Genetically Engineered Food,* a catalogue of foods sold in grocery stores throughout Canada, specifying which contain and do not contain GE ingredients.

Friends of the Earth Canada

- Developed *GMOs Out of Our Food and the Environment* (Friends of the Earth, 2004b), a document that outlines national and international efforts to pressure governments and gain public support to ensure adequate testing and labelling of all Genetically Modified Organism (GMO) products, restrict the sale and inclusion of GMOs in food aid, make the biotech industry liable for GMO pollution, and ban GMO rice and terminator seed technology.
- Maintains pressure on governments, including Canada's, to ratify the Cartegena Protocol on Biosafety, an international agreement to regulate trade in GMOs.
- Published several studies examining GMOs and the companies and organizations promoting them.
- Examines in its Global Environmental Justice campaign (Friends of the Earth, 2004a), the inequitable wealth, consumption of resources, and creation of pollution by industrialized northern nations and the deleterious impacts of this on poorer countries.
- Investigates how trade agreements, corporate practices, and global development aid supports inequities. As a part of this campaign, talks about the injustice experienced by migrant farm workers in developing nations who are paid low wages, experience dangerous working conditions, and are poisoned by pesticides used on crops destined for northern markets.

David Suzuki Foundation

- As a part of its sustainability program, composed a policy document, *Sustainability within a Generation,* which includes nine goals, a few of which relate to food and agriculture indirectly, such as "shifting to clean energy," and some more explicitly, such as, "to produce healthy food" (Boyd, 2004).
- In its discussion of food and agriculture, focuses mostly on the reduction of pesticide use and a broad-based move to organic farming in Canada.
- Recommends encouraging connections between local producers and consumers, phasing out some and reducing the use of other pesticides,

eliminating tax breaks and introducing additional taxes on pesticide purchases, making labelling of foods mandatory, reducing agricultural run-off, encouraging low or no till practices, increasing riparian belts, and encouraging a healthy diet of reduced meat consumption (David Suzuki Foundation, *n.d.*).

- Put together the report *The Food We Eat: An International Comparison of Pesticide Regulations.*

Sierra Club of Canada

- Its Health and Environment area includes the program Safe Food and Agriculture which focuses on educating rural and urban citizens about the environmental impacts of agriculture, lobbying the federal government on agricultural policy and taxation, encouraging the preservation of publicly funded agricultural research and small-scale family farms, and increasing urban gardens (Sierra Club of Canada, *n.d.*b).
- Developed specific campaigns, including on biotechnology; food irradiation; intensive livestock operations; and mussel, shrimp, and salmon agriculture.
- Highlights information on fair trade.
- Discusses a more inclusive view of the food system/chain in comparison to the other ENGOs. For example, in its overview of fair trade, it explains that "when we think about the significance of safe food, it is important to remember that food safety and health issues go beyond our individual bodies . . . health can and should include a myriad of ideas, from the health of neighbors, to the health of ecosystems, to the health of those who we may not know but to whom we are linked through the global chain that brings food to our plates" (Sierra Club of Canada, *n.d.*a).

Canadian Environmental Network

- Has an agriculture caucus, the mandate of which is to work "towards an agricultural future which respects and nourishes both human communities and the natural world, and is truly sustainable environmentally, economically and socially" (Canadian Environmental Network, *n.d.*a).
- Chose as the theme of its 2006 annual conference agriculture and the environment, and based on the forum discussion is producing the document *Green Paper on Agriculture and the Environment: Challenges and Solutions.* Topics covered in the draft of this report include biodiversity (habitat loss, crop plant and livestock genetic diversity, GMOs), water consumption, pollution (sediments, nutrients, pesticides, and

veterinary drugs), soil erosion (organic matter, chemical contamination, organic farming versus conservation tillage), and air (greenhouse gases, energy from biomass, ammonia, and pesticides).
- Like the Sierra Club, provides online discussion of many of the dimensions of the food chain and the connecting threads between national and international aspects of this issue, as well as acknowledging not only the ways that current agricultural practices are contributing to the deterioration of the rural environment but also how they provide environmental and social benefits.

All five of these national environmental groups demonstrate recognition of some of the ecological and social connections of many of the environmental issues around which they have organized their agro-food-related campaigns. Each has identified specific agricultural practices, for example biotechnology/GMOs, as environmental and social concerns, and the Sierra Club and the Canadian Environmental Network have extended this analysis to provide a more encompassing examination of many aspects of the industrial agro-food chain. Although most of these environmental groups have generally not explicitly framed agricultural and food issues as matters of environmental justice (the Friends of the Earth being the exception), they engage in varying degrees of discussion about issues of inequity in the distribution of wealth and resources, and the imposition of "potentially dangerous" foods on poorer nations around the globe. For example, Friends of the Earth's Global Environmental Justice (2004a) and GMOs Out of Our Food and the Environment (2004b) campaigns both highlight how the North consumes an inequitable share of the world's resources, produces the majority of the pollution, and imposes its technologies on the South with little regard for these southern citizens' rights, communities, health, or long-term development. Likewise, the Sierra Club's Fair Trade campaign discursively weaves together threads of connection between the North and the South, and the rich and poor in its discussion of health extending beyond our individual bodies to include consideration of the health of the ecosystem, our neighbours, and even people we don't know but to whom we are linked through the global food chain (Sierra Club, *n.d.*a).

Yet at the same time, the examination of the overall system of production, distribution, sale, and consumption of food products Gottlieb insists is necessary is largely missing, and related discussion to matters of justice has not been systematically engaged. For example, the Canadian Environmental Network's draft *Green Paper on Agriculture and the Environment* provides a fairly thorough investigation of the multiple dimensions of the eco-agro complex in its examination of biodiversity, water, soil, and air, yet it limits the spatial analysis to Canada, and aside from a few brief mentions of the processing, storage, and distribution of food, focuses almost exclusively on the production aspects of agro-food sustainability (Castro, 2006). In this discussion of sustainable agriculture, issues of equity, class, and distribution are pushed into the background, if mentioned at all, almost as if enhancing environmental sustainability was an end in itself (Clancy, 1994). In contrast, Greenpeace and the Suzuki Foundation are much more consumer oriented: Greenpeace recommends a diet free of GE foods, and the Suzuki Foundation insists on mandatory labelling, eating less meat, and connecting producers and consumers. The Suzuki Foundation engages with issues that directly affect producers, such as low or no till cultivation, increased riparian belts, decreased agricultural run-off, organic farming, and a reduction in the use of pesticides. Missing, however, is discussion of how these recommendations might adversely affect many family agricultural operations in Canada and throughout the world, or how consumer demands for cheap food play a role in supporting the industrialized agricultural model. In the end, in each case, there are significant social, ecological, and equity links in the agro-food system missing in these discussions.

A more deliberate environmental justice perspective could highlight areas of these campaigns that are contradictory, incomplete, and not as ethical as they might first seem. The insistence on an embedded, multiscalar view of injustice encourages a socio-geographic view of the agro-food system in which the agro-ecological specificities of the farm are connected to the quality, price, and availability of the food, and in which the asymmetries of power, consolidation, and corporatization of the agricultural and food production, processing, and distribution sectors are connected to the lives of those who produce, process, and consume that food. To illustrate a couple of ways this form of analysis might be applied, I'll examine the recommendations these environmental groups make for ethical consumerism and discuss some of the contradictions, tensions, and matters of justice and equity that arise in thinking through the environmental justice dimensions of the agro-food system.

Some Environmental Justice Dimensions of the Canadian Agro-Food System

One of the ways in which each of the five national environmental groups in Canada addresses environmental injustice in the agro-food sector is by promoting ethical consumerism through wise food choices, be it consuming food that is local, organic, wild, or GE free. Examples include Greenpeace's shopper's guide to avoiding GE food (Greenpeace, 2006) and the Sierra Club encouraging "consumers to make food choices which are consistent with their values ... to buy locally, to buy organic products when possible, to avoid farmed fish and meat raised in factory farms and to limit the distance from farm to fork . . ." (Sierra Club, *n.d.*a). Such ethical consumerism is at one level a strategy for expressing social and environmental values: through personal consumption one can satisfy multiple goals – the desire for healthy and pleasurable food, and support for environmental well-being and social harmony (Johnston, 2005). Moreover, because in contemporary society our identities are so often focused on consumer choice and lifestyle, food and food consumption are integrated into our personal, social, and cultural ideals (Sandler, 2005; Soper, 2004). Building on this, slow food advocates have argued that food choices can be more than culinary hedonism; they can be entry points into larger political projects that engage with issues of social equality and environmental sustainability (Leitch, 2003).

Although frequently seen through the lens of individual rational decision making, commodity choices are nevertheless embedded in capitalist power relations marked by biases of gender, race, class, and ethnicity, those very social dimensions environmental justice has succeeded in highlighting but which too often are pushed into the background in discussions of sustainable agriculture and healthy, accessible food. With respect to ethical consumerism, this often translates into food choices such as wild salmon, organic produce, and GMO-free foods being elite commodity fare – healthy yet expensive foods that frequently depend on a system of marginalized labour, and especially in Canada are grown out of place and out of season and then shipped thousands of kilometres to upscale food markets (Guthman, 2003). Despite these foods having lost some of their "green gleam," the organics sector continues to grow dramatically. Research shows that the majority of this sector is largely composed of individuals who are more concerned about personal health than they are about environmental sustainability and social justice. Hence, health concerns are privileged over social concerns, and environmental issues away from the field are largely ignored (Dimitri and Oberholtzer, 2006; Macey, 2004). In turn, this continual growth in organics has not only attracted but been promoted by corporate interests, as conventional industrialized food companies such as Dole have gotten into the business of organics, and in so doing helped transform large portions of this food sector into large-scale, long-distance, and monoculture food production (Jaffe, Kloppenburg Jr., and Monroy, 2004; Delind, 2006). Clearly, we have a situation in which privileged eating has intrinsically led to impoverished eating on many levels (Guthman, 2003).

While organics transform into big business, on the other end of the neoliberal food provisioning spectrum there has been a continual increase in and institutionalization of food banks throughout Canada. There is a food bank in every province and territory, as well as a progressive increase in clients over the quarter-century since the first food bank was set up in

Edmonton (CAFB, 2006). Such contradictions illustrate the kind of culinary "pleasures" that result from systemic maldistribution, as the corporate food sector continues to try to consolidate and maximize profits.

Enter onto the food stage local food. The Sierra Club proclaims the need to "buy locally and wherever possible, directly from farmers" (Sierra Club, *n.d.*c), the Canadian Environmental Network recommends adopting and promoting the food miles concept, which it explains involves "monitoring and minimizing the distance food travels from field to fork (Canadian Environmental Network, *n.d.*b), and the Suzuki Foundation's proposed policy encourages local food connections between producers and consumers (Boyd, 2004). It appears that promoting locally grown foods has become a clarion call for ethically responsible food consumers. As Kloppenburg, Hendrickson, and Stevenson (1996) explain, "the distance from which [the] food comes represents [consumers'] separation from knowledge of how and by whom what they consume is produced, processed and transported . . . how can they understand the implications of their own participation in the global food system when those processes are located elsewhere and so obscured from them? How can they act responsibly and effectively for change if they do not understand how the food system works and their own role within it?" (p. 34). In this sense, local food is thought to be an antidote to the global corporate food regime and its reliance on intensified energy inputs, product durability, and homogenization. As such, local food is posited as counter-hegemony to the failures and constraints of globalization.

Yet, as DuPuis and Goodman (2005) argue, the "representation of the local and its constructs – quality, embeddedness, trust, care – privilege certain analytical categories and trajectories whose effect is to naturalize and occlude the politics of the local" (p. 361). By virtue of the label, someone has to decide what constitutes local food, and in this act make decisions about the inclusion of particular people, places, and ways of life. DuPuis and Goodman assert that "such unreflexive localism can lead to potentially undemocratic, unrepresentative and defensive militant particularism" (p. 362) that can feed notions of purity and perfection through self-regulation and pit one region against another. Moreover, rather than contesting industrial agriculture, such activities may reinforce the programs, techniques, and procedures that are integral to market rule, productivism, and global competition (DuPuis and Goodman, 2005).

In a Canadian context, such recommendations also fail to account for the agro-ecological realities of food production in this country. Although growing produce for the consumer market is possible in many regions of the country, this is a seasonal and limited form of agricultural production. The majority of Canada's farmland, 82 percent, is located on the Prairies (Alberta, Saskatchewan, Manitoba), which experiences a temperate climate and short growing season and which primarily produces grain, oilseeds, and beef for an export market (Agriculture and Agri-Food Canada, 2001). Saskatchewan and Manitoba are also two of the more sparsely populated provinces, and even combined with Alberta, the population is not large enough to consume all the food produced on the Prairies.

As local food strategies in Europe demonstrate, the concern for quality and regional provisioning can support multifunctional agriculture and in this way serve as a powerful political force against global industrial agriculture (Leitch, 2003). However, many of these initiatives are not transferable across national, cultural, and ecological contexts. What is required is a rethinking of the local across spatial scales – a multiscalar analysis that pays attention to local institutional interests and equity in the context of global forces and pressures. Agricultural policies and practices embedded in a politics in place that counter the homogenized techniques and placelessness of capitalist industrial agriculture are required, but these will not occur alone through calculating the distance food has travelled before arriving on a consumer's plate. As DuPuis and Goodman (2005) explain, an alternative agricultural movement that is responsive to the needs of the environment and people where it is located will arise through "an inclusive and reflexive politics in place [that] understand[s] local food systems not as local 'resistance' against a global capitalist logic but as a mutually constitutive, imperfect, political process in which the local and the global make each other on an everyday basis" (p. 369).

More generally, although ethical consumerism can create pressure to alter specific industrial agro-food practices (e.g., dolphin-friendly tuna), the question still remains of what possibility these consumer choices have to fundamentally challenge global industrial agriculture. As Johnston (2005) explains, the consumer sovereignty ideals that underlie ethical consumerism are based on the belief that the pursuit of self-interest leads to greater public well-being, thereby valorizing individual choice over collective action by organized citizens and the state to combat social and environmental problems. In this context, social change and citizenship are conceived in exceptionally narrow terms: action invoked through the market place involving a choice between brands. Such limited democratic participation fails to provide or encourage engagement in citizen-based activities such as lobbying politicians for improved working conditions, environmental regulations, or fair trade; a systematic review of the food system; or reducing consumption.

Hence, the creation of a just and environmentally sustainable public commons requires more than simply consumers seeking to maximize individual pleasure and articulate their personal identities. As Johnston (2005) puts it, "Public spheres require democratically engaged citizens who consider not just their purchasing power, but their civic rights and responsibilities to a public good existing beyond the realm of individual purchases" (pp. 23-24). As the recent transformations in organics illustrate, too often consumers overlook citizenship-based rights such as safe working conditions, a decent wage, and environmental sustainability in order to claim their sovereign right to product choice. This does not suggest that buying local, organic, wild, or GMO-free food is either right or wrong. Rather, the argument is that in the context of a capitalist industrial agro-food system, the application of a fixed label cannot always assure social justice or environmental sustainability, for such consumer signs are inherently unstable and open to corporate co-optation. In a globalized and deregulated agro-food system, generalizations about food as a commodity have always been uneven and contradictory. Therefore, understanding the specific dimensions of the ways in which these contradictory relations manifest in a particular place requires clarification and analysis of the distinctiveness of the socio-cultural and political patterns of internationalization in different agro-food sectors and their regional hinterlands (Watts and Goodman, 1997).

Conclusion

Canadian national environmental groups in their discussions of agro-food issues have to varying degrees engaged with matters of social justice and environmental sustainability. In particular, Friends of the Earth, Sierra Club, and the Canadian Environmental Network illustrate quite diversified agro-food strategies that both explicitly and implicitly engage with the notion of environmental justice. However, their discussions of environmentally sound farming practices or ethical consumer food choices, though important work, are socially and environmentally incomplete. An analysis of their ethical consumerism campaigns illustrates significant oversights they make in failing to examine in-depth complex agro-food system chains.

The multiscalar and socially complex formulations of Canadian environmental justice scholars such as Debbané and Keil, Draper and Mitchell, and Teelucksingh provide a useful frame of analysis for tracing the power geometrics and multiple spatialities central to the agro-food complex. Applied to food studies, this framework highlights how working toward systemic change in the agro-food system will not occur through a focus on single issue topics. Moreover, given the current political and economic organization of agriculture, the purpose of which is to primarily produce profit and only incidentally to produce food, ignoring the attribution of cause in an analysis of the entire food chain seems foolhardy in the development of an alternative agro-food system. At some level, efforts have to be directed at identifying and transforming the structures, values, and processes that have given rise to unsustainable and unjust aspects of agricultural production, distribution, and consumption (Sumner, 2005).

Notes

1 In using the term *broad-based environmental organizations,* I am referring to environmental groups that work on a diversity of environmental issues, from air and water pollution to saving habitat to reducing and controlling toxic substances. I am delineating these environmental groups from national environmental organizations such as the World Wildlife Fund or the Canadian Parks and Wilderness Society, which are focused almost exclusively on the protection and establishment of parks, green spaces, and wildlife habitat.

2 Each year the Canadian Association of Food Banks conducts a survey on hunger in Canada based on the use of food banks throughout the country over a one-month period. It recognizes that although this does not capture the full extent of hunger in Canada, it does give some measure of general trends. In 2006, over 753,000 people faced hunger (CAFB, 2006).

References

Agriculture and Agri-Food Canada. 2001. Profile of Production Trends and Environmental Issues in Canada's Agriculture and Agri-Food Sector. http://www.agr.c.ca/cal/epub/1938e/1938-0021_e.html.

Agyeman, J. 2005. *Sustainable Communities and the Challenge of Environmental Justice.* New York: New York University Press.

Amin, A. 2002. Spatialities of Globalization. *Environment and Planning A, 34*(3), 385-99.

Boyd, D.R. 2004. Sustainability within a Generation: A New Vision for Canada. David Suzuki Foundation. http://www.davidsuzuki.org/files/WOL/DSF-GG-En-Final.pdf.

Brulle, R.J., and D.N. Pellow. 2005. The Future of Environmental Justice Movements. In D.N. Pellow and R.J. Brulle (Eds.), *Power, Justice, and the Environment: A Critical Appraisal of the Environmental Justice Movement* (pp. 293-300). Cambridge, MA: MIT Press.

Buttel, F.H. 1997. Some Observations on Agri-Food and the Future of Agricultural Sustainability Movements. In D. Goodman and M.J. Watts (Eds.), *Globalising Food: Agrarian Questions and Global Restructuring* (pp. 344-65). New York: Routledge.

CAFB (Canadian Association of Food Banks). 2006. HungerCount 2006. http://www.cafb-acba.ca/documents/HungerCount_2006_EN_Summary_ WEB.pdf.

Canadian Environmental Network. *n.d.*a. Agriculture Caucus. http://www.cen-rce.org/eng/caucuses/agriculture/index.html.

–. *n.d.*b. RCEN Agriculture Caucus Workplan. http://www.cen-rce.org/eng/caucuses/agriculture/index.html.

Castro, M. 2006. *RCEN Green Paper on Agriculture and the Environment.* Ottawa: Canadian Environmental Network. http://www.cen-rce.org/eng/projects/2006conference/Green%20Paper%20final%20version%20feb%202007%20-%20e.pdf.

CBC News. 2006. Breast Cancer More Common in Farm Workers: Study. http://www.cbc.ca/ health/story/2006/10/12/breastcancer-farm.html.

Choinière, Y., and J. Munroe. 1997. Farm Workers Health Problems Related to Air Quality Inside Livestock Barns. Factsheet. Toronto: Queen's Printer for Ontario. http://www.omafra.gov.on.ca/english/ livestock/swine/facts/93-003.htm.

Clancy, K. 1994. Social Justice and Sustainable Agriculture: Moving beyond Theory. *Agriculture and Human Values, 11*(4), 77-83.

David Suzuki Foundation. *n.d.* http://www.davidsuzuki.org/.

Debbané, A.-M., and R. Keil. 2004. Multiple Disconnections: Environmental Justice and Urban Water in Canada and South Africa. *Space and Polity, 8*(2), 209-25.

Delind, L.B. 2006. Of Bodies, Place, and Culture: Re-Situating Local Food. *Journal of Agricultural and Environmental Ethics, 19*(2), 121-46.

Dimitri, C., and L. Oberholtzer. 2006. A Brief Retrospective on the U.S. Organic Sector: 1997 and 2003. *Crop Management* doi:10.1094/CM-2006-0921-07-PS. http://www.plantmanagementnetwork.org/pub/cm/symposium/organics/Dimitri/.

Draper, D., and B. Mitchell. 2001. Environmental Justice Considerations in Canada. *Canadian Geographer, 45*(1), 93-98.

DuPuis, E.M., and D. Goodman. 2005. Should We Go "Home" to Eat? Toward a Reflexive Politics of Localism. *Journal of Rural Studies, 21*(3), 359-71.
Friends of the Earth Canada. 2004a. Global Environmental Justice, International Program. http://www.foecanada.org/index.php?option=content&task=category§ionid=9&id=23&Itemid=113.
–. 2004b. GMOs Out of Our Food and the Environment. http://www.foecanada.org/index.php?option=content&task=view&id=122&Itemid=93.
Gottlieb, R. 2005. *Forcing the Spring: The Transformation of the American Environmental Movement.* Washington, DC/Covelo, CA: Island Press.
Gottlieb, R., and A. Fisher. 1996. Community Food Security and Environmental Justice: Searching for a Common Discourse. *Agriculture and Human Values, 13*(3), 23-32.
Greenpeace Canada. 2006. Say No to Genetic Engineering. http://www.greenpeace.org/canada/en/campaigns/ge.
Guthman, J. 2003. Fast Food/Organic Food: Reflexive Tastes and the Making of "Yuppie Chow." *Social and Cultural Geography, 4*(1), 45-58.
Hanson, L. 2007. Environmental Justice across the Rural Canadian Prairies: Agricultural Restructuring, Seed Production and the Farm Crisis. *Local Environment, 12*(6), 599-612.
Jaffe, D., J.R. Kloppenburg Jr., and M.B. Monroy. 2004. Bringing the "Moral Charge" Home: Fair Trade within the North and within the South. *Rural Sociology, 69*(2), 169-96.
Johnston, J. 2005. Shopping for Justice: Ethical Consumerism and the Case of the Whole Foods Market. Paper presented at the joint meeting of the Agriculture, Food and Human Values Society and the Association for the Study of Food and Society, Portland.
Kloppenburg, J. Jr., J. Hendrickson, and G.W. Stevenson. 1996. Coming in to the Foodshed. *Agriculture and Human Values, 13*(3), 33-42.
Leitch, A. 2003. Slow Foods and the Politics of Pork: Italian Food and European Identity. *Ethnos, 68*(4), 437-62.
Macey, A. 2004. "Certified Organic": The Status of the Canadian Organic Market in 2003. http://www.agr.gc.ca/misb/hort/org-bio/pdf/ OrganicsREPORT2003_e.pdf.
McDonald, M.A., L. Arragutainaq, and Z. Novalinga. 1997. *Voices from the Bay: Traditional Ecological Knowledge of Inuit and Cree in the Hudson Bay Bioregion.* Ottawa: Canadian Arctic Resources Committee and Environmental Committee of Sanikiluaq.
Nestle, M. 2002. *Food Politics: How the Food Industry Influences Nutrition and Health.* Berkeley/Los Angeles: University of California.
Pollan, M. 2006. *The Omnivore's Dilemma: A Natural History of Four Meals.* New York: Penguin.
Sandler, R. 2005. Review of *The Ethics of Food: A Reader for the 21st Century,* by Gregory Pence, ed. *Journal of Agricultural and Environmental Ethics, 18*(1), 85-93.
Schlosser, E. 2001. *Fast Food Nation: The Dark Side of the All-American Meal.* Boston: Houghton Mifflin.
Sierra Club of Canada. *n.d.*a. Fair Trade. http://www.sierraclub.ca/national/programs/health-environment/food-agriculture/campaign.shtml?x=558.
–. *n.d.*b. Safe Food and Sustainable Agriculture. http://www.sierraclub.ca/national/programs/health-environment/food-agriculture/index.shtml.
–. *n.d.*c. What Is Organic Food? http://www.sierraclub.ca/national/programs/health-environment/food-agriculture/organic-food.shtml.
Soper, K. 2004. Rethinking the Good Life: Consumer as Citizen. *Capitalism Nature Socialism, 15*(3), 111-16.
Spurlock, M. (Director). 2004. *Super Size Me* [DVD]. New York: Showtime Networks.
Sumner, J. 2005. Value Wars in the New Periphery: Sustainability, Rural Communities and Agriculture. *Agriculture and Human Values, 22*(3), 303-12.
Sustainable Table (Producer). 2003. *The Meatrix* [Online video]. New York: Free Range Studios for the Global Resource Action Center for the Environment (GRACE).
Teelucksingh, C. 2002. Spatiality and Environmental Justice in Parkdale (Toronto). *Ethnologies, 24*(1), 119-41.

Walker, G., and H. Bulkeley. 2006. Geographies of Environmental Justice. *Geoforum, 37*(5), 655-59.
Watts, M., and D. Goodman. 1997. Agrarian Questions: Global Appetite, Local Metabolism: Nature, Culture, and Industry in Fin-De-Siècle Agro-Food Systems. In D. Goodman and M.J. Watts (Eds.), *Globalising Food: Agrarian Questions and Global Restructuring* (pp. 1-32). New York: Routledge.
Wekerle, G.R. 2004. Food Justice Movements: Policy, Planning, and Networks. *Journal of Planning Education and Research, 23*(4), 378-86.
Wilkins, J., and M. Eames-Sheavly. 2003. A Primer on Community Food Systems: Linking Food, Nutrition and Agriculture. *Discovering the Food System: An Experiential Program for Young and Inquiring Minds.* http://foodsys.cce.cornell.edu/index.html.

12

Rethinking "Green" Multicultural Strategies

Beenash Jafri

> After two decades of protest against the allocation of toxic wastes sites in predominantly poor black and aboriginal communities, American environmental justice advocates can claim significant wins . . . but in Canada, there has been comparatively little activity to report . . . the absence of discussion may be attributed to the wider reluctance to engage in "race talk" in Canada. Racism is considered an ugly word here, and mention of it often puts many people on defence.
>
> – Andil Gosine, "Talking Race," p. 3

It is indeed very difficult to talk race in Canada, and this is one possible reason why an environmental justice movement has not taken force here to the extent that it has in the United States. It is not, however, very difficult to "talk multicultural"; in Canada, multiculturalism is official policy and a central component of the Canadian national identity. It is a policy based on representational practices that reinforce the mythology of Canada as a white settler nation. Through narratives that script indigenous peoples as dead, dying, or disappeared, and peoples of colour as perpetual newcomers, white people are centred as the gatekeepers, the primary and original inhabitants and rightful owners of the nation (Razack, 2002, p. 5). As many scholars have pointed out (e.g., Bannerji, 2000; Das Gupta, 1999; Thobani, 2007; Mackey, 2002), the official policy of multiculturalism defines Canada as the legacy of its two founding nations – English and French – while paying homage to indigenous peoples, viewed as virtually extinct and celebrated superficially through signifiers such as totem poles, canoes, and various other cultural artefacts, and welcoming in newcomers (primarily people of colour) who after even several generations of settlement are asked to explain "where they are from." Indigenous peoples and people of colour in this narrative are simultaneously inside and outside; they always remain external to the white nation, yet essential to its existence both symbolically and materially (Bannerji, 2000, p. 65).[1]

In recent years, several environmental groups (particularly in Toronto) have taken up multiculturalism within their programming. These organizations have employed strategies based on notions of multiculturalism in an attempt to "diversify" their practices or to outreach to immigrants of colour. I refer to these here as "green" multicultural strategies. Using data from research carried out from 2002 to 2003, I critically examine in this chapter examples of multicultural outreach projects and programs carried out by three environmental organizations in Toronto – Greenest City's Multicultural Greening Program, the Toronto and Region Conservation Authority's Community Development for Multicultural Environmental Stewardship Project, and the Evergreen Foundation's Urban Oasis Project – asking the question, how have the politics of multiculturalism been played out in Canadian environmental groups?[2] In undertaking this analysis, I hope to problematize conventional approaches taken by environmental groups in Canada to address the homogeneity of their predominantly white, middle-class membership. By providing a critical lens and context, I aim to open a space for the development of strategies that are more conducive to radical, transformative, and systemic social change.

This chapter is divided into four sections. I begin by introducing the three organizations and their multicultural outreach projects. In the next section, I look at the ways in which multiculturalism was taken up in these projects. In the third section, I examine some of the lessons to be learned through these projects regarding alternative approaches to environmentalism, namely via environmental justice frameworks. Finally, I look at how the context of organizational structures and funding politics influenced the sorts of questions, issues, and ideas explored within the projects. In the Conclusion I consider some of the central themes of the chapter and implications for environmentalists.

"Green" Multiculturalism in Practice: Three Organizations

Three organizational projects were examined in this study: Greenest City's Multicultural Greening Program, the Toronto and Region Conservation Authority's Community Development for Multicultural Environmental Stewardship Project, and the Evergreen Foundation's Urban Oasis Project. They are profiled briefly below.

Greenest City – Multicultural Greening Program

Greenest City was originally affiliated with FoodShare Toronto as one of three Green Communities Canada initiatives focused on air-quality improvement and greenhouse-gas reduction. However, with program funding cuts by the Harris government in Ontario, it was established as an independent organization in 1996. According to an organizational information sheet, "Greenest City's values include justice, diversity, participation, and community, informing how we work and with whom" (Greenest City, *n.d.*b). Community, grassroots participation, and healthy environments provide the framework for much of the organization's programming, which focuses on local Toronto communities. Greenest City's Multicultural Greening Project (MCGP) began initially as the Chinese Community Outreach Project in 1996. In 1999, the project expanded to look at other communities and other issues relevant to (immigrant) communities of colour in the city. In 2002, it was placed on hiatus.

Toronto and Region Conservation Authority – Community Development for Multicultural Environmental Stewardship Project

The Toronto and Region Conservation Authority (TRCA) is a fairly large organization with a nontraditional management structure. It was founded in the 1950s, following Hurricane Hazel, with an initial mandate for flood control that was later broadened to include issues of water quality, wildlife, and vegetation, and most recently, to include people, as stewardship and education have become key components of the organization. Unlike most NGOs, which may be structured and governed via political boundaries, the TRCA is organized around a watershed system (there are nine watersheds in the Greater Toronto Area), each of which has its own staff and strategic plan. A board of directors, however, with representatives from political jurisdictions, also helps guide the work of the organization. According to one staff member, pre-1995 the TRCA was a provincially funded body, but it has now obtained charitable status in order to apply for private funding grants. The Community Development for Multicultural Environmental Stewardship (CDMES) Project was initiated in 1997 to address the lack of representation of people of colour at events and on committees and boards, which was of concern especially in light of the Greater Toronto Area's highly multicultural population.

Evergreen Foundation – Urban Oasis Project

The Evergreen Foundation, founded in 1991, is a national environmental NGO with "a mandate to bring nature back to our cities through naturalization projects" (Evergreen Foundation, *n.d.*). The main interests of the foundation, according to a staff member, include biodiversity, native species (flora and fauna), habitat conservation, urban land trusts, school ground naturalization, outdoor education, and bicycle-use promotion. Evergreen follows a standard NGO structure, with a board of directors, permanent and contract staff, and volunteers. It is relatively well staffed and financed, and at the time of research was receiving funds through the City of Toronto, the TRCA, several private foundations, and corporate sponsors, including Toyota, Wrigley Canada, and the TD Friends of the Environment

Foundation. The Urban Oasis Project (formerly known as Where Edges Meet) was launched in Toronto in 1998 to bring nature to inner city neighbourhoods (Evergreen Foundation, 2002). The overall goal of the project was to build "stronger, healthier communities by engaging neighbourhoods in the creation of sustainable community ecological gardens within City parks and, in the process, [foster] a strong sense of stewardship and pride" (Lewis and Irvine, 2001, p. 2).[3]

Critiquing "Green" Multicultural Strategies

In this chapter I review two "green" multicultural strategies. The first is the use of the framework of diversity as a starting point for the multicultural outreach projects undertaken by Greenest City, the TRCA, and the Evergreen Foundation. The second is the linking of environmental stewardship with Canadian citizenship and inclusion.

Diversity versus Anti-Racism

> [Diversity is] a goal . . . we try, but aren't completely diverse, because the environmental movement is affiliated with white people, and we're always on the outside, trying to get into communities of colour.
>
> – Greenest City staff member

> Urban Oasis is very much aimed at getting new Canadians involved in learning about environmental issues and natural history . . . St. Jamestown [is] the prime example, a community that receives new Canadians, there are always new immigrants and it's a high-rise population so people have no opportunities to be (a) exposed to and (b) involved in environmental issues, so the project is an opportunity to be involved and exposed to areas of Canadian life such as the urban recycling program.
>
> – Evergreen staff member

An issue common to the three organizations profiled here, as well as to many environmental organizations more generally, is a preference for frameworks based on diversity or multiculturalism, as opposed to anti-racism. The deployment of these sorts of frameworks, which in general focus on cultural difference as the primary gap or problem limiting the participation of people of colour in the environmental movement, serves to obfuscate both the role of systems of power and privilege in creating this exclusion (e.g., Philip, 1992), as well as the ways in which frameworks of diversity and multiculturalism may themselves reinforce white hegemony (e.g., Bannerji, 2000; Mackey, 2002; Razack, 2002; Thobani, 2007).

Each organization began its respective projects from the premise that

Toronto was multicultural, and that its central goal therefore was to broaden the scope and outreach of its programming to reach "diverse" communities. Meanwhile, there was limited analysis or awareness of whiteness or white privilege as a system of power and the ways in which this was potentially being structurally and discursively reinforced within organizational practices. Such an analysis might have produced an entirely different approach to understanding and addressing the predominantly white composition of each respective organization – one, for example, that might entail an organizational change process involving an examination and restructuring of organizational goals and priorities, decision-making practices, employment policies, programming choices, and so on.

Indeed, within the three organizations surveyed, there appeared to be great interest in promoting racial harmony but not in examining the structural roots leading to a lack of harmony. It followed that several research participants asserted a preference for the term *multicultural* over *anti-racism* because it was viewed as more positive and all-encompassing. *Anti-racism,* meanwhile, was identified as more negative and less amenable to community building. With the exception of one staff member at Greenest City, none of the participants identified multiculturalism as a government policy or the ways in which it might be an ineffective mechanism for addressing institutionalized racism faced by people of colour. Not surprisingly, there appeared to be little consensus or collective understanding within the organizations surveyed of how race and racism might be understood. Greenest City perhaps came the closest to addressing this issue through its very general and liberal "diversity statement," which provided a framework for the way in which the organization operates.[4]

Commonsensical understandings of race, racism, and multiculturalism were reflected in the multicultural outreach projects undertaken by these organizations. For the most part, each organization emphasized perceived cultural barriers (e.g., language, lack of information, lack of communication) rather than structural or discursive racism as the primary factor leading to the exclusion of people of colour from its environmental work. For example, Greenest City's MCGP project began as the Chinese Community Outreach project in 1996 and was based on the premise that people of colour in Toronto were excluded from environmental programming because of language and cultural barriers, and the size and density of Toronto's Chinese community (Greenest City, 1998). Similarly, at the TRCA, the emphasis of the CDMES Project was on increasing participation of people from "diverse cultural backgrounds" in the organization and on eliminating barriers to participation.[5] The strategies suggested for elimination of these barriers were primarily better communication and education, rather than advocacy for structural change. As one TRCA staff member commented, the only way in which the stewardship and education work she did was necessarily "multicultural" was through the provision of bus tokens, translation, childcare, and so on to increase accessibility to the events; there was nothing particularly unique about the content and approach of the programming. Essentially, the work appeared to be about presenting the same messages in an accessible and relevant way.

Evergreen's project, although not quite so explicit in its goals to engage people of colour or immigrants in organizational activities (e.g., the title of the project, Urban Oasis, had a generic feel), suggested these intentions through the use of language such as *diversity, inclusion,* and *celebration* in project materials.[6] For instance, one project report asserted that "working in highly multicultural neighbourhoods allows cultural diversity to meet with natural diversity" (Lewis and Irvine, 2001, p. 2). Although the organization's approach, with its emphasis on providing "opportunities" for new immigrants to be involved in the environmental movement rather than on eliminating barriers to participation, differed slightly from that of the TRCA and Greenest City, it was similarly focused on promoting the same environmental messages and priorities to a different audience.

Inclusion through Environmental Stewardship

> When people come to Canada, they arrive with a dream, an understanding about this country and assumptions about the service providers. We believe that in order to get new Canadians integrated into the Canadian system, an understanding of our environment is as important as understanding the social or economic system.
>
> – Chandra Sharma,
> *Human Connections,* p .1.

> Working side by side with their neighbours to create and tend native wildflower gardens, the project provides a venue for people who are socially isolated, often due to language barriers or for health reasons (e.g., lack of mobility) to meet members of their community in a safe and accessible environment. It also gives people with minimal access to naturalized green space, many of them new Canadians, a chance to garden with native plants and learn about Canada's natural heritage.
>
> – Anne Marie Lewis and Sheana Irvine,
> *Gardening as a Catalyst for Community Health,* p. 1

A second and related green multicultural strategy that emerged in this research is the linking of citizenship discourses with environmental stewardship. In the cases of the TRCA and Evergreen, environmental stewardship was perceived to be a (neutral) catalyst for facilitating citizenship and inclusion in Canadian society. Accordingly, both organizations made presentations to newcomers at the Language Instruction for Newcomers to Canada classes (offered by the federal government) as an outreach strategy. A TRCA staff member noted that these presentations were an important way to "catch people in the first five years" in Canada and to educate them about environmental issues in Toronto.

There are two general ways that we might problematize this strategy. First, by neutralizing the notion of Canadian citizenship and nationhood – that is, by removing it from its social and political context – existing relations of power based on white supremacy (and the subsequent marginalization of indigenous peoples and people of colour within the nation) are reinforced. As described earlier, discourses of Canadian nationhood simultaneously centre and invisibilize white settler subjectivity. Second, the idea that newcomers need to be introduced to environmental stewardship or need to be educated about the environment also obscures the ways in which people experience the environment and environmental stewardship in everyday ways – in our homes, bodies, workplaces, neighbourhoods, and so on. As Dorceta E. Taylor (1993) argues, people of colour have remained outside environmental movements "because of the way in which environmental issues were framed, the kinds of issues that were focused on, and the ways those issues were strategically addressed" (p. 58).

As is evident through the discussion in this section, "green" multicultural strategies can be challenged not only in terms of their inability to address the systemic exclusion of people of colour within the environmental movement but also in terms of the reinforcement of relations of domination and subordination.

Rethinking "Environmentalism"

> White, middle-class people have more time; they can hire nannies . . . and, of course, you have more time to take care of the environment when your basic needs are met.
>
> – Greenest City staff member

In the process of doing their multicultural outreach work, Greenest City and TRCA staff found themselves questioning the strategies they had put to use; there are important lessons that emerged from this work.

TRCA – Toward New Strategies?

At the TRCA, project workers found that conventional approaches to promoting environmental stewardship did not seem to work with immigrants of colour. For instance, although she was not aware of any of the research or action on environmental racism through the environmental justice movement, a TRCA staff member remarked that issues such as water quality and waste management – those that had a more direct impact on people's daily lives – seemed to garner more interest and attention in CDMES activities than issues such as wildlife preservation and conservation. In essence, she discovered through her work and experiences what environmental justice activists and scholars have been reiterating for over two decades (e.g., Bullard, 1993). Another TRCA staff member had similar insights that hinted toward the need to rethink environmentalisms, though her proposed strategy did not necessarily accomplish this. She noted that "what I discovered is that it's not true that [people of colour] are not interested [in environmental issues]. The big problem is that the environment is perceived in a different sense. In the first five years in Canada you're looking for jobs, housing – you're not going to volunteer spare time to plant trees . . . you have to make it value-added for people trying to settle here, you can't expect the environment to be big on their agenda."

At the time this research was conducted, however, these learnings from TRCA staff members did not appear to lead to significant shifts in programming. As will be discussed later in this chapter, this lack of organizational change not only was linked to the analysis brought by individual staff, but was the result of systemic factors.

Greenest City – Possibilities for Environmental Justice Programming

Meanwhile, at Greenest City, self-reflection actually led to the evolution of an environmental justice-focused project as the limitations of the Chinese Community Outreach Project became apparent. Although in technical terms, the project had been a success – all of the project objectives were achieved, gardening and tree planting were identified as activities that interested the community, and the project built links with other community organizations and associations and eventually expanded to work with different communities – the multicultural aspect of the project was called into question. For example, a question raised by staff was, what was special about a MCGP community garden project identified as multicultural, when most community gardens in Toronto could have been identified as such?

As a result, the need to work from an environmental justice perspective and to think about how communities were affected by urban spaces became a critical next step (Greenest City, 1998). A second project coordinator was hired in 1999 to research and test ways to expand the mandate of the MCGP. The objectives of this revised project included testing the following:

(1) The feasibility of working with communities not necessarily united or self-identified by a specific geographic relationship (as in the case of working with residents of an apartment building where the land or the space/object of a particular greening action is obvious, i.e., commun-

ity gardening on the land adjacent to the building) but perhaps united or self-identified by ethnicity or neighbourhood or issue.

(2) The feasibility of working in partnership with existing groups that express a desire to undertake projects to improve their local environment, their relationship with it, and the sustenance it could provide them.

(3) The feasibility of community social, economic, and environmental development under these conditions. (Greenest City, 1998, p. 2)

The focus now moved away from helping communities of colour participate in the environmental movement to taking a nontraditional approach to environmental issues, one that explicitly brought to the fore concerns of livelihood and empowerment, and, moreover linked together the generally separate spheres of environment, society, and economy. This approach did not fall outside Greenest City's mandate; the organization already took a broad approach to the environment and environmentalism, understood to be "everything" – the rural, the urban, the social, the economic, and the political.

Interestingly, the MCGP was not framed as an explicitly environmental project. A former project coordinator remarked, "My role wasn't advocacy . . . but work that was doing more hands-on [activities] and engaging the public [who] don't necessarily think about doing environmental things . . . it wasn't packaged [to suggest] that we were doing environmental work." She also commented on the racialized nature of the environmental movement, drawing attention to the need to better contextualize environmental issues, noting that the movement "is not inclusive because the way that it's being presented is a very Canadian way . . . the people active in it are from a Canadian background, so unless you can get people to bridge the gap it's hard to [include people from] different races."

Thus, while Greenest City did not explicitly set out to undertake an antiracism project, because of the emphasis on urban space, socio-economic inequity, access, and distribution, racialized injustice (i.e., racism) could easily be brought to the table. As one staff member noted, because poverty is racialized in Canada, and because Greenest City was interested in working with those people and communities facing the most barriers and marginalization, its target groups were "automatically low-income immigrants, by default."

The Multicultural Greening Project provides an example of how a more integrated analysis of race, space, and class might result in an entirely different approach to environmental issues that reconceptualizes mainstream notions of the environment and environmentalism. However, the work could not be sustained. In the following section, I explore some of the underlying issues behind this.

Organizational Politics and the Politics of Funding

It is critical to contextualize the discussion in this chapter in terms of organizational structures and funding politics. The projects undertaken by the three organizations were not merely the choices of individual staff, board members, or volunteers but were produced systemically within the context of these politics. As many ad hoc grassroots groups discover when they go the route of institutionalization (often in order to access more funds and resources, to gain legitimacy, or to sustain projects and programs), increased bureaucracy often leads to a decrease in autonomy and capacity to implement agendas for radical change (e.g., see INCITE! Color of Violence, 2007). For example, the charitable status of any given organization can be revoked and funding pulled or reduced if the organization is perceived to be overly politicized or radical (e.g., Ford-Smith, 1997, p. 231).

In the case of the three organizations, there was much more room for dialogue, debate, and organic growth within Greenest City – as a relatively small NGO (at the time of research, there were only four paid employees) with a community oriented focus – than was possible at the Evergreen Foundation, a much larger organization receiving both public and private funding. Meanwhile, staff at the TRCA – an organization that is predominantly white, male, and middle class – shared that it was difficult working on a small project of the TRCA separate from the rest of the organization. Their suggestion was for the CDMES to be integrated into the TRCA as a key component of the organization; at the time this research was conducted, the program was not given high priority and unlike some of the TRCA's other programs did not receive core funding (aside from the provision of the project coordinator's salary). Because of this, fundraising had to be done through outside sources, and the program could be carried out only for a short time before funding had to again be sought. Thus, from 1997 to 2001, funding for the CDMES came from the Trillium Foundation and Environment Canada; since 2001 funders have included Environment Canada and TD Friends of the Environment Foundation, and in 2003 a proposal was being processed to request funding for another two years.

Funding politics have also supported the absence of environmental justice work within environmental organizations in Canada. While organizations seek out funding to fulfill their mandates for education, advocacy, service provision, and so on, the requirements of funding programs can, ironically enough, produce conditions that limit or diminish organizational capacities not only to do their work but to grow, expand, and challenge themselves (though many organizations do frequently develop strategies that attempt to challenge or work around these limitations). Narrow, highly specific funding criteria based on compartmentalized, noncritical understandings of racism and environmentalism reinforce "business as usual" and a lack of any integrated analysis of racism within environmental organizations. Moreover, output-based funding program objectives discourage meaningful change processes from taking place while encouraging the diversion of organizational resources toward the administrative labour that becomes necessary to maintain funding within this model (e.g., Ng, 1990).

For example, as a staff member at Greenest City related, there had been ample discussion within the organization about how to contextualize environmental issues within the rubric of social justice, equity, and anti-oppression. In particular, the Multicultural Greening Program encouraged a great deal of self-reflection and raised many questions about the framing and conceptualization of environmental issues in the organization. However, because slow, long-term change and processes cannot be quantified into quotas or concrete, measurable outcomes – which are preferred by grant-giving institutions – the project was placed on hiatus after six years, and Greenest City (based on its small size and limited resources) made the decision to limit its focus to issues of air quality and smog.[7]

Although organizational structures and funding politics complicate possibilities for radical transformative change to take place within environmental organizations, this is neither a sign to deter this work nor to simply continue with existing approaches. Rather, it poses critical questions for environmental organizations interested in redefining and reconstructing environmentalism: How might these politics be navigated? What strategies might need to be undertaken in order to subvert, resist, and challenge mechanisms that limit the capacity of environmental organizations to engage in anti-racist change?

Conclusion

> [With respect to] environmentalists who have become more alert to the influence of social constructions of culture in organizing and experiencing environmental conditions, and concerned about the homogeneity of their movements . . . their work is not to merely bring people of colour into their fold, but to engage people of colour – and a commitment to anti-racism – towards the creation of whole new objectives and tasks.
>
> – Andil Gosine, "Race-Racism in Popular Environmentalism," p. 193

In this chapter, I have problematized green multicultural approaches and in doing so have demonstrated some of the ways in which the environmental movement is not a neutral space but, in fact, a site for the reinforcement and maintenance of relations of power. The examination of projects by Greenest City, the Toronto and Region Conservation Authority, and the Evergreen Foundation are concrete examples of this. Some of the experiences of staff working on these projects, moreover, provide important lessons for environmental groups looking to imagine alternative anti-racist approaches that are potentially more radical and subversive, and which push the envelope in terms of what we understand environmental work to be. This work is challenging; if environmental organizations want to bring people of colour into their membership and programming, a more concerted effort to examine and address all of the ways in which this exclusion is maintained – toward the development of new agendas for change – is required. Ultimately, this will mean looking beyond the issue of the exclusion of people of colour within the environmental movement – which is a symptom of broader sets of concerns, rather than a problem in and of itself – to the multifarious ways in which environmental organizing strategies might be supporting relations of domination and subordination.

Notes

1 This, however, does not imply that the marginalization of indigenous peoples and people of colour within the nation is congruent with one another. See, for example, B. Lawrence and E. Dua, "Decolonizing Antiracism," *Social Justice, 32*(4), 120-43.

2 This research involved interviews with six key informants who had worked or were working on these projects in the selected organizations, as well as a review of project and organizational literature. This chapter is a modified version of the author's senior honours work thesis, completed at the Faculty of Environmental Studies at York University in 2003.

3 Specific project objectives included to engage members of culturally and economically diverse communities in sharing ideas, experience, and knowledge in order to improve the quality of the local natural environment; to involve other environmental and community organizations offering complementary programs in the establishment of a stronger, more diverse program framework for achieving results; to build community capacity by recognizing and building on existing strengths, skills, and assets of the community partners and to help develop new skills and knowledge through environmental education and training in restoration techniques; and to raise awareness at the community level about the concepts of habitat loss and restoration, biodiversity conservation, and the role of naturalization in creating environmental literacy (Evergreen Foundation, 2002).

4 According to its website in 2003, "Greenest City recognizes that human and ecological communities are inherently diverse. Diversity is essential to the health of communities and the environment. We seek to understand, respect and celebrate diversity. Therefore, we: 1. Engage varied communities in projects that unify people towards common goals; 2. Develop approaches that are appropriate, varied, and innovative. We recognize that one model does not fit all; 3. Work to ensure that Greenest City's decision-making bodies, staff and volunteers reflect the diversity of the communities we work with." (Greenest City, *n.d.*a). This statement no longer appears on Greenest City's website. However, as of 23 August 2007, embracing diversity is included among the organization's values: "We view diversity as a key source of the creativity needed to generate integrated solutions to today's environmental challenges. We increase our impact on these challenges by working with and for the diverse individuals and communities that make up the City of Toronto" (Greenest City, *n.d.*a)

5 Barriers, which were identified by the TRCA in conjunction with community leaders and anti-racism consultants, included language; access, availability, and transportation; lack of contact with environmental advocates; lack of relevance of environmental projects to people's well-being; lack of information and opportunity; not knowing how to con-

tribute; lack of resources, both financial and organizational; and lack of environmental education and awareness in a culturally specific way (Sharma, 2001, p. 2).

6 Specific project goals included encouraging people from diverse cultural backgrounds to take an active role and responsibility in the environmental management of their communities; building partnerships with various multicultural groups, agencies, and residents in the planning and implementation of environmental plans that reflect the specific needs of the community; identifying and eliminating barriers that limit the participation of new Canadian communities in environmental management activities; developing skills in facilitation, ecological restoration, and site planning; increasing opportunities for employment; and changing habits and practices that benefit our watershed ecosystem (Sharma, 2001, p. 2).

7 There are two general reasons grant-giving institutions prefer quotas or measurable outcomes. First, grants tend to be allocated for relatively short periods (i.e., under five years) on a project-by-project basis. Funding institutions tend to demand, therefore, that tangible results are achieved within that time frame. Second, funding institutions are held accountable to their stakeholders (e.g., individuals, families, corporations, or other institutions in the case of private foundations; Canadian taxpayers in the case of public funding programs) and need to demonstrate that their finances have been distributed as per the mandate and objectives of a given funding institution.

References

Bannerji, H. 2000. *The Dark Side of the Nation: Essays on Multiculturalism, Nationalism and Gender.* Toronto: Canadian Scholars' Press.

Bullard, R.D. 1993. *Confronting Environmental Racism: Voices from the Grassroots.* Boston: South End Press.

Das Gupta, T. 1999. The Politics of Multiculturalism: Immigrant Women and the Canadian State. In E. Dua and A. Robertson (Eds.), *Scratching the Surface: Canadian Anti-Racist Feminist Thought* (pp. 187-205). Toronto: Women's Press.

Evergreen Foundation. 2002. Where Edges Meet: Celebrating Cultural and Ecological Diversity. 2001 report to the Toronto Community Foundation.

–. *n.d.* About Us. Evergreen Foundation. http://www.evergreen.ca.

Ford-Smith, H. 1997. Ring Ding in a Tight Corner: Sistren, Collective Democracy, and the Organization of Cultural Production. In M.J. Alexander and C.T. Mohanty (Eds.), *Feminist Genealogies, Colonial Legacies, Democratic Futures* (pp. 213-58). New York and London: Routledge.

Gosine, A. 2003. Talking Race. (Editorial.) *Alternatives, 29*(1), 3.

–. 2002. Race-Racism in Popular Environmentalism (PhD dissertation, York University, Toronto).

Greenest City. 1998. Multicultural Greening Project: Background and Job Description for Feasibility Study . . . Project Development and Community Outreach Worker. http://www.greenestcity.net/11.html

–. *n.d.*a. About Us. http://www.greenestcity.net.

–. *n.d.*b. Greenest City. Information sheet. http://www.greenestcity.net/2.html.

Incite! Color of Violence. 2007. *The Revolution Will Not Be Funded: Beyond the Non-Profit Industrial Complex.* Boston: South End Press.

Lewis, A.-M., and S. Irvine. 2001. Gardening as a Catalyst for Community Health: Where Edges Meet; Celebrating Cultural and Ecological Diversity/An Evergreen Project. Paper presented at the conference of the Federation of Ontario Naturalists, *Woods Talk: Community Action to Conserve Ontario's Woodlands,* Toronto.

Mackey, E. 2002. *The House of Difference: Cultural Politics and National Identity in Canada.* Toronto: University of Toronto Press.

Ng, R. 1990. State Funding to a Community Employment Center: Implications for Working with Immigrant Women. In R. Ng (Ed.), *Community Organization and the Canadian State* (pp. 165-83). Toronto: Garamond Press.

Philip, N. 1992. *Frontiers: Selected Essays and Writings on Racism and Culture.* Toronto: Mercury Press.
Razack, S. 2002. Introduction: When Place Becomes Race. In S. Razack (Ed.), *Race, Space and the Law: Unmapping a White Settler Society* (pp. 1-20). Toronto: Between the Lines.
Sharma, C. 2001. Human Connections: Multiculturalism and the Environment. Paper presented at the Sustainability Network seminar Communicating Environmental Messages in a Diverse Society, Toronto.
Taylor, D.E. 1993. Environmentalism and the Politics of Inclusion. In Robert D. Bullard (Ed.), *Confronting Environmental Racism: Voices from the Grassroots* (pp. 53-61). Boston: South End Press.
Thobani, S. 2007. *Exalted Subjects: Studies in the Making of Race and Nation in Canada.* Toronto: University of Toronto Press.

13
Coyote and Raven Talk about Environmental Justice

Pat O'Riley and Peter Cole

speaking for ourselves hmm
so where's the howling and yelping and thrackkkking and rapping
seems to me there's a whole lot of english language ideas and words
getting in the way of what is really going on
and who's really being impacted by capitalism consumerism and cronyism
from c to shining c

how many times have you had to change dens coyote in the last 3 months
let alone the past 100 years and all of your relatives too
how many of the trees where my ancestors have lived for centuries
are gone now turned into paper in the name of progress
toilet paper newspaper parliamentary memo paper *lo stesso vecchio sempre*
but I guess our homes and lives don't count
our historicies of contact don't matter
what about the tree nations who have been contract killed for centuries
first by merchants of discovery then by empire and now
in the name of market forces and the majority (white and other nonaboriginal) interest
is toilet paper really a step up from trees and bark and leaves and
what kind of carbon sink is toilet paper or the eatons
(oops I mean sears) catalogue for that matter
they don't respire or photosynthesize or produce oxygen or take up CO_2
they don't provide shade they don't keep soil from blowing away
they are not a home for lichens mosses mushrooms birds insects
how much development is sufficient is sustainable or will development go on
and on and on how high a gnp do we need
what does "enough" mean in a capitalist society
"until the next buy or sale comes along"

stroke one stroke one stroke one enough
stroke one stroke one stroke one enough
drum drumdrum drum drumdrum drum drumdrum

you know raven these human beings are a greedy lot
they want more than their fair share they want everybody's share
they want to make up all the rules including the rules of the natural world
damming blasting clearcutting wiping out whole species
how did they become like that is greed contagious
and what is the epidemiology of consumerism
kill everything destroy everything in the name of economics
in the name of wreckonomics

when laws are passed in provincial legislatures or in the house of commons
who asks our sisters and brothers the animal insect bird worm moss grass fungus
and tree nations how they feel about it they're the ones who are most directly affected
they're the ones most in need of environmental justice
most indigenous societies had respectful relationships with the natural world
because they/we had foresight they looked ahead seven generations
to their great grandchildren's great great grandchildren for a start
after all if the process is not sustainable in the long run neither are we

first peoples did not practice development it was not part of our ethic
we did not have real estate or investment portfolios stock and bond markets
we were not infected with consumerism capitalism
we had to wait for the immigrants to bring over their first [sic] world ideas all hail

voiceover: it has come to our attention that human beings are unnatural
and their mistakes are being compounded daily by stupid acts of worldwide vandalism
against the rest of nature therefore they will all be turned into salt poooof!
fiat nacl not good for the ailing heart

where does environmental justice for one species meld into biocide for/against others

killing everything in the name of the economy   corporate
profits   gst   taxes
shareholder satisfaction   *comfort*
what other point of view is prevalent
what other context is pervasive save the hypercapitalist one
does it take place upstream downstream slipstream slidestream
from the landed money   *pl op! go the weasels the fishers the otters*
is their absence their extinction merely an assemblage
of diverse and necessary silences
people have become reliant on "conveniences"
piled upon conveniences   and cannot despite their own desire to
stop the avalanche of consumeropathy   I want   I need I lack!

you know raven these academics writing down ideas
glueing printed words to a spine on one side
organizing them into phrases clauses paragraphs chapters and margins
it never seems to amount to much except the same old same old
university libraries are infirmaries filled with unwell words
mortuaries agape with clichés *o verba* I mourn for you
for what you have done to the world through the people
you have molded   *o populi* I weep for you
for your stupidity   for not taking care with the tools
that shape you petrified   unchanging

imagine   sending your own mother
to the hospital to be cared for and they turn her into a fossil overnight
uh ms mr x your mother is uh   ready to be picked up now
we have turned her body and life into words theories assumptions
prognoses
well uh no she's not actually that well she's dead in fact
yes unfortunately   another *iatrogenic misdemeanor*
we should have left her alone   instead we multiplied (our greed)
and our need   and she was caught in the middle   la mia mama
in the middle of what?
uh in the middle of a very productive oilfield forest fishery
she just   got in the way   oh mother earth   I grieve for you!
or   oh here's your mother back she's okay
she just doesn't have any vital functions left or signs   sorry
and by the way we eviscerated her harvested her organs   next please!

raven maybe you have to hill the words   like you do potatoes
or thin out the ideas as you do when peppersandtomatoescarrots get too
crowded

or perhaps you could companion plant like with the three
  sisters cornbeanssquashshsh
that does wonders for crop yield
it's not crop yield we're after   in this instance
it's getting that book we've been working on since time immemorial
it's coming down to the crunch   cash crunch ab crunch crispy crunch
maybe you should companion plant some ideas for a change
ha remember that whistle from captain crunch cereal
that you could blow and make free long distance calls

but what are you doing getting caught up in this business
you know better than that

well   you know sometimes you just have to

revive a 500 year old headache earache noseache bellyache
oooow it hurts everywhere   except most of the time I couldn't tell
because I thought that's what life felt like
the pain is a constant   companion

no no no   sometimes you have to contribute

to what   the mayhem   the mayhaps
is that what they call it these days "contribute"
I say give 'em a good bite or a good scare
claw their hair   rake their scalp   that'll keep 'em on their toes
or at least off their buts sic

rather than on ours

up thompson way nor-man region they call it   northern manitoba
there used to be a trap line these cree people had
but then valeINCO   (uh INCO   *vale*!)
found nickel and   goodbye trapline

cruelty thy name is

now in the middle of where white society assumes no cree people ever
  were
is thriving thompson manitoba
with aboriginal people occupying the streets
outside the post office kitty corner to the rcmp hq

across from the library   the canadian tire parking lot
run down hotels   bars with seeds in them
"thompson: the hub of the north"   yes that's one of the pubs
at city hall they talk about *them* as *we've got a problem downtown*

hmm well if INCO hired some of those aboriginal people
took a chance   like they do with white canada
maybe they wouldn't be on the street
if someone cared to train them   enhance their skill levels
maybe they wouldn't be hanging around the bars
if you live on the street long enough it's where you feel you deserve to be
INCO thompson INCO sudbury   helllooo!

reprise (re-take) where have all the first nations gone long time passing
if indians are so invisible when it comes to (not) employing them
how is it that they are such a visible "problem" when they are
  unemployed homeless
sleeping in the single person's hostel the magic triangle
hostel   bar   rcmp building
or maybe   hostel   vendors   (be/hind/side the post office) post office
that would be isosceles   all angles acute butnotso *sowieso*

when they ask you for money on the street (which they rarely do)
  remember
they are your landlord your landlady
talk like this in thompson and you'll be ko'd pretty quick
you see   biiig money is to be made at INCO by mostly white investors
and nobody's going to want to waste high salaries on no'count indians
pardon me but does INCO participate in the federal contractor's equity
  program
does valeINCO do business (besides behind the scenes   *in camera*
  (obscura))
with the federal government
and who   is   INCO   oh a corporation a person
soooo shareholders who don't live in thompson determine policy
and INCO is legally bound to make money and not squander it   oh!
I see   how is it that the canadian government (name one)
allows this tyrrany to take place   oversight?   supervision? *untermensch*
you've got it   the feds are the cheerleaders   with their pockets stuffed
with shit   I mean money but it becomes shit after a while
end of lesson 1   catabola (a hybrid of catabolism and parabola) for the
  millions

howso parabola

locus of points in a plane equidistant from a given point/focus
which is the native people each and all equidistant from jobs at INCO

so mathematically you can graph the racism I mean the lacunae in fair
treatment

chorus *are we blue are we blue we are too telling you we are blue*
acknowledgments to grant clarke
why are we blue raven we got mcpromises and mcjobs
and mchouses and mcaccess to mcwelfare eat your heart out tim
mchorton

the trapdoor below the proscenium arch opens and closes
coyote hangs in midair unphased the smell of nickel pervades
and stephen harper's cologne augh oil of olé
welfare is part of environmental injustice in canada
more after a word from our sponsor the international nickel emporium

environ mental justice mm has a nice sound to it (paws)
but it doesn't smell very good most academic words don't have a long
shelf life
and would be better off in the compost before publication actually
save all the trees the forests of words replacing real forests

yes too many words getting in the way of not enough action

but words are action coyote blahblahblahblahblah say it backwards now
look at the royal commission on aboriginal peoples
ha they should have saved themselves the trouble
and given the money for high quality first nations housing and education
there's a misnomer aboriginal people already knew
what the problems were RCAP just gave the feds time
to make up more indian genocidal policies

canadians are so apathetic! they sit on their but(t)s
and watch tv shows about genocide in their own country
and say somebody should do something about it
and isn't it a shame and they change channels
then vote in (or don't vote at all) another liberal or conservative
government
or a right of centre ndp government

well come on down! *we are down*

I think that if we want to make a difference we have to colonize
the english language create centres of power
minorizing the majority discourse gilles deleuze called it
we have to empower our speaking and writing and thinking
otherwise we are wasting our time

empower okay word! fetch! fetch! come on now! there you
go no action
words are the cats of the symbol world

but words good words can empassion us to act to care about others
they can encourage us to advocate on behalf of others
that's what stories are all about
and besides people have to act on their *own* behalf behaves behavesnot
even though you need a critical mass of gumption of inspiration of
doittoit
of get up and go before it's got up and left
before self empowerment clicks in
it's not like magic *poof!! you are now self empowered end of struggle*

but people forget things touch their hearts
then they get up go to the fridge and find some snack
end of throughline beginning of middle of linethrough

I see what you mean how many hear the call or heed the messages
look at the global protests the same small group of protestors
and undercover cops in the faces or pocketbooks
of the politicians and the paramilitary
did you see the other day the rcmp and qpp troublemakers
who had infiltrated the antiglobalization movement at montebello
they were the ones caught throwing stones
trying to incite the peaceful protestors to riot
the police break the law to keep the peace
and then they say on national television that they have to do that
a kind of pre-emptive strike where did they learn that tactic from?

honi soit soit honi ah comme la pluie a plû

you know how linda smith talks about research
being a dirty word to many first peoples

well how about the word manitobahydro or hydroquebec bchydro
ontariohydro
say those words in those provinces to aboriginal people
and first you'll hear an audible intake of breath followed by an uneasy
silence
or many uneasy silences even retroactive silences if you care to discern
but the good old cbc they're on our side right they're exposing
the crap right?
the good old cbc where were they when . . . and . . . and . . . and . . . ?
they are on the side of "a good story" no matter how they contort
the aboriginal side of the story
how come cbc is never there when the injustice is in full swing
batter up WTO's on first *transition music and that's the kind of day*
it's been (for indians) thanks lloyd! kevin peter

raven uphome I used to be able to drink out of the river eat the plants
live off (and as) the land but now aye there's the rub "but now"
I don't even dare eat a goose or mouse or a muskrat
they're toxic sinks just as we are have become
bioaccumulative sinks bathtubs swimmingpools
now I come down to stanley park high park and find a well fed
minicanine
which has been fed healthy food that's how I survive CHOMP!
colonizing the colonizers I'd even settle for a semicolon
or an unguarded comma

speaking of toxins coyote and environmental justice
how about the millions of people in canada talking on cell phones all
over the place
scaring all the birds who can hear microwaves confusing us
"yeh like I mean you know like yeh s/he's cool
like you know I uh really dig her/him like you know"
just a minute I got a call on the other line
oops I think I hit a moose or a schoolchild or a
bicyclist bumpbump splossssh
yes well maybe talking on the phone and driving don't mix that well
right like maybe make one the "thing" and the other the chaser
it's a form of verbal auditory pollution it inundates my ears my senses
we need a talk dam uh rather than damn talk
yes coyote did you have anything to do with the fact that birds hear
microwaves
did you plan this as a young pup knowing that cell phones

(and who knows what's coming in the future) would be developed
by the twoleggeds

me no on the other hand it's also a great gift
if birds like you raven can hear microwaves then you can
probably hear aliens trying to communicate with us
from across the universe

aliens yes that's the right term

well raven I think all versions of so-called environmental protection
are useless the old versions and the young versions
have no teeth not even a mouth no oral masticative apparatus
except for an already bunioned corned and blistered alveolar ridge
it's a sham a façade it's business first last and everywhere else

it's soooo vague any measures to protect the environment in canada
issues of human health and wellbeing always come after "the economy"
consumerism is the god and we eat the god and poop the god
isn't it all so grand

I mean like when will the politicians stop prioritizing the next election
over everything else maintaining power just for the sake of ego we go
they go

that's what capitalism is sister money always comes first
human nonhuman or otherthanhuman life be damned
it's all about prostrating ourselves before the corporate citizenry
we have all bought into it and are buying into it
that is what sustainability means to the government universities
corporations
it means money comes first
sustainability is a way of saying let's pretend we're doing something
good
for the environment something noble *so are we all all honourable men
schsch*
sustainability should be about giving back more than we take
wakey wake
otherwise there is no sustainability there is only taking taking
and if you add it all up you've subtracted or added negative numbers

remember how we were so critical of the internet
and how it's making children and adults fat and lazy

well the internet has helped to spread ideas of environmental justice
it has helped to bring together and connect civil society groups
yes well dams have created the internet
maybe we should only use watts and amps that come from wind power
solar power
for our computers for our networking
otherwise we are building foundations of justice on histories of injustice

aaaaah

hmmm here comes raven's big "but"

but racism justice inequity it's all about power
and aboriginal people have been antéd in with the immigrants
(which is everyone else)
and issues we have lived with for 500 years are going to have to wait in
line
with all of the new inequities the inequities of the visitors the
newcomers
how is it that our sovereignty is not seen to be a priority
by a caring national government and dam(n) what the "people" think
of our manifestos so what if the majority don't like them
isn't that why we have the charter of rights and freedoms
so that the tyrrany of the many the ignorance of the mostly
does not trample on rights of the minority and of the first peoples

are you sure it's about waiting in line raven perhaps we could wait
together
I'm a good waiter most creatures of the "wiiiild" are
most savage beasts need patience
it is the stuff of life survival
personally I've been waiting for capitalism and its minions to go back
to where they came from waiting and bioaccumulating

hey! I'm talking here okay well I guess I was interrupting
but not if you think about priorities
will it be the priorities of the newcomers again that are forefronted
foregrounded ie the mainstream racial groups
(who have always ignored us in their conversations about racism)
the race is on
we've endured terra nullius lingua nullius historia nullius indiana nullius
and now a new cashier opens up and we're still at the back of the line

pardon me but we are also in a bit of a predicament here
and there and *un peu partout se:kon boozho tansi ama7 sqit*
uh we're being genocided here
please take a number an indian number a social security number
I'm sorry the rest of the cashiers are on lunch
there are only so many of us

ah there's that "only" word it's a dangerous one
it means that now we have to stand in line behind ourselves
you know once the last indian has died everyone will say
"well we were getting to you but you didn't last long enough
you should have gone to customer service"
customer service!? none of them were indians and wouldn't even waste spit
besides nobody would let us in the doorway let alone past it
when it comes to indians you can talk around the block
and every time you get back home none of the talk
makes any difference nothing changes it's word weather again word climate
academic words and legal definitions and legislated laws are made of straw
(itself a word)
nobody enforces anything relating to aboriginal rights which freezes up the courts which is pretty much
what the object is to have the complexion of justice
which has no roots and blows away in the next public or corporate uproar
about how indians should pull up their socks or darn them
and the westcoast ones should put a retread on their burks
without the substance of justice or assumption that laws will be enforced
because they aren't popular they're too just toward us
and well everybody knows the attitude of the paramilitary and the military
in canada to aboriginal people but not better than those of us
who have been abused by the injustice system
indians know about the moonlight tours they're popular with the police
the rc's in not just saskatoon but across this great haff kaff country
they take them off inco road in thompson or up past the railroad station
damn troublemaking indians slap leather slap slap and when you run
out of bullets you've always got a taser or two

environmental justice for indians! what universe are you living in
you don't measure up your thinking machine is seized up

measure? is somebody baking a cake or frying bannock
I like it when you can just throw in a handful of what nonconsideration
a dash of irresponsibility a pincenez of fluff
place on a greased panopticon keep in the fridge
until the power runs out and black mold
covers everything and even then even the molds starve to death
from waiting for dogot who spent so much of her/his life
as a museum specimen that s/he forgot that she was more
than just white words and academic ditheration
granma! ah my little one ah!

excuse me I'm going out for a smoke I can't stand this acamumbojumbo

but you don't smoke

I'm going to smoke somebody or something
this tok tic tok tic blatherskate mutherdrubbing
it's a retroactive virus I do not wish to have an intimate relationship with
never end a sentence in a
proposition would you like to kill chaos theory with me
I'm deputizing you maybe we'll pick off postcolonialism while we're at it
it has no roots either except in academic confabs
so it blows around like sage brush

I could use some sage right about now

m and a brush

sssssssh there's an environmental justice sustainability conference going
on
will somebody run that by me again put it into a blender first
add some probiotics antibiotics postbiotics and call me in the morning
coyote you have to remember to stop interrupting
interrupting? 500 years of silence
with a little rant here a little rant there
here a rant there a rant everywhere a rantrant
surely it's allowed uh by the way who is the allower
please point out that author/ity figure

raven is work green?

huh?

is capitalism green what is the colour of capitalism
and environmental justice is it green
and what colour is environmental racism or is it outdated now
is it too controvertial so they're toning it down
taking out the tawnies and the shawnees
but especially the local first nations they're the real troublemakers
always asking for rights I ask you what rights do you have
you have the indian act isn't that enough so enough already
you have your justice I have mine you're standing in my sun

excuse me I thought they were distributing justice today
but it's only newcomers distributing to newcomers
with all indians either permanently on deck or in the hole
pardon meeeee uh we've been in redistribution centres for quite some time now
and we'd like to go home uh yes kitsilano stanley park ville marie byward market and yorkville yes before "development" took place
my people were buried there for a thousand generations
and that's not counting before the iceages

my turn to a tree and I do not claim to be speaking for trees
but as an advocate to a tree there are no laws or thoughts or justices
there are only loggers chainsaws grapple chains chompermachines
pine beetles and acid rain there is either standing leaning or lying down
being a tree is being a home
but now there is the option of two by four sawdust chipboard and shiplap

who wants to be shiplap? how about 3/4 inch plywood good one side
hmmmm not too many takers paper anyone want to be 20 pound
excuse me I'll have some valium please ritalin!
they've clearcut my medicine grove and patented everything
except for those who have copyrighted and trademarked

excuse me (I keep saying this!) hi you can't see me because I'm still
in the spirit world
I was supposed to be born in 1955 but my mother was forcibly sterilized
at the charles camsell federal hospital in edmonton because she was an indian

and the doctors were given (w)orders to do what's best for us
I'd like to be born   *us too us too*

raven what does environmental justice mean to you

aw are we getting back to that   nothing!

nothing?

no   nothing

then you're not part of the conversation

no coyote I haven't been cited by academics
I am in an ntz a no talk zone permanently
who gets quoted gets legitimized and we as nonacademics
get shot at or ignored or chainsawed deaths in custody? mostly indians

they are trying to tell us that environmental justice is a new theory
they're just getting around to thinking about it
because environmental injustice has been ignored especially
environmental injustice on the poor racialized people indigenous
  peoples

but things are different in canada right?
no it's all about immigrants and immigrants
and immigrants and immigrants of all colours all citing each other
trying to badmouth us by saying that we too
are immigrants that we are transberingial refugees
or by putting forward theories that have nothing to do with indians
but everything to do with western thought westernized thunk
we are conceptual appendages of the western imaginary
in the end environmental justice is a western conversation
a western conceptualization it has nothing to do with us
because it's newcomers versus indians
my relatives came over with the mayflower
like I said newcomers versus indians
hmmm

academic spaces including those relating to aboriginal peoples
are about deaboriginalizing space and meaning and experience
and filling them with newcomer racializational rhetoric and aboutness
  googolplexed

how to binarize without essentializing
well maybe a few essentials are needed and at least
a parsing through use of binary (aboriginal-nonaboriginal)
until we can get that one jump started and running on all cylinders
auricle ventrical whoooosh auricle ventrical whoooosh
other/wise aboriginal people will yet again be mixed in with the
caesar dressing with all the bacon and anchovies and cream
until there is no taste left to us
as anything more than part of a western discourse

jr;;p fpvypt zoyjoml zo mrrf dp,r jr;[
pardon me?
oh pardon me! the white person who was trying to write about me
got hisher fingers off the homerow keys and was writing gibberish
but since you ask
I am having a bit of a problem you see I've become a trope
and I'm not sure what to do about it I have become the written "other"
trapped in words expressions assumptions by mainstream experts
about about about sounds like a frog with homophonitis

uh not to worry take two similes and call me in the morning
and uh maybe have your dna tested you might have a retrovirus
you mean like the 1950s
no more like the 1550s
you have been likely infected with words bigtime
and they have been knitted or darned or goddammed into your genes
nothing to worry about just uh stay away from westerns for a while
and professional sports mascots

uh and another thing I'm not "an aboriginal" I hope
that I have moved past at least the adjectival stage
as well as the adjective stage though I don't wish to be a noun
an english trope I certainly don't wish to spend eternity (another noun)
as a mere descriptor or modifier it would actually be nice to be/come a verb
though verb itself is a noun not to worry I still have *ucwalmicwts kalanwi!*

ah so you want to be "an aboriginal *person*"! no! too many syllables

help help
what's the matter?
I'm being interpreted

I think I'm trapped in laughin reruns
and another thing
yes
when academics talk about me they're not talking about me at all
but about themselves about their search for evidence white academic evidence
well how about a little experience next time you're in the great hereafter
and you're thinking about a future role try my mocassins size 43 high arch

raven I'd like to say something

go ahead

I've just done a study and discovered that all 600+ reserves in canada
are examples of environmental racism as perpetrated as sustained by the federal government the colonial government and the citizens
just thought I'd add that a little bit of experience a little bit of hearsay
oh did I say that? I mean oral testimony I know it's a bit of a stretch youdon'tsay
yep that's what it's like where I'm from
yep that's what johnnie said
hmmmm hearsay *par ouidire*
inadmissible in court add out

mommy
yes dear
I think I've become a toxic waste site
that's nice dear have some rezwater
maybe you'll feel better
mommy
yes dear
what's that stuff floating in the jello?
environment it's the environment dear

excuse me EXCUSE ME yes I have a cottage on the georgian bay
well not really a cottage it costs about $3 million
anyway I am a victim of environmental injustice
all the foul air from toronto hamilton sudbury pennsylvania and the ohio valley
has settled over the whole area and a lot of millionaires are getting angry
well why don't you just call your mp or the nyse and launch a complaint
next caller?
hi my baby sister came out weird well I mean uh different

and we only ate fish from the river
what river dear?
the st lawrence (name your river) we only had fish 3 times last month
sorry to hear that you might want to buy fish from the
supermarket next?

you know what's really ironic raven is that work as constructed by the
immigrants
is put forward as an unassailable ethic a *conditio sine qua non*
"if you don't work you're a bum" and bums are a no no
and people who work in blue colour jobs like miners and factory workers
often have to live in the immediate vicinity of the effluents which they
produce
and people whose source of food is "wild" food also suffer from ingesting
poisons
which exist in their sources of nutrition
so the health of the individual and of the society is at best secondary
to the health of the economy that is the health of corporations
western society is based on gambling the stock market is the basis
for capitalism and cutting back is the euphemism for
giving in to corporate interests
being bought off sound familiar s t e p h e n! paul! j e a n!
b r i a n! . . .ringo
anybody home?

is there a real separation of corporate and state interests
are governments political arms of corporations
and even if people are the electors of governments
how informed are they how much apathy do they have
how much resignation do we all have as to "how things are"
how lazy have we become?
and what is the level of responsibility of those who do not vote?
of those who choose to remain ignorant?

raven environmental justice is just an expression
it is meaningless unless people actually do something about
what it purportedly is
and if apathy is the main response then it will not change
it will accelerate like the indian act and the only passengers
not given seatbelts/shoulderharnesses are the indians
and the next collision should be taking place right about NOW!

buuuut if someone interfered with say sports teams or alcohol consumption
there would be lots of passion even compassion rather than apathy
being the mainstream relationship with the living world

people just don't care

weeeeell I don't know if I'd go quite that far

not many seem to care
and perhaps that is our job
to care to lead by example who has the energy! multitasking is us

it's all so discouraging raven genocides happen mass rapes child murders
they are happening today and people are too busy
with their everyday problems (sound of page turning channel flipping fridge opening lawn mower car changing gears beer can opening)

can people wake up or will it always be the job of the minority
to say hey wake up stupid and smell the crap
you're eating it you're putting it on and into your children

environmental justice if I don't take a stand I have no place
from which to criticize others
but is writing about it actually doing something and if all
anybody did was write about it would anything get done?
I don't think so

what are you saying exactly

me I'm just talking I don't have any answers
because I don't believe in problems

I'm going to fly over the people shopping at walmart and poop on their suv's

and I'm going to go hunting for organic food in the park
locally produced

here!

what are those?

noise cancelling earphones   they cancel out the sound of
  microwaves infrared waves

ahhhhh!

kukwstum   skennen

Contributors

Julian Agyeman is Professor and Chair of Urban and Environmental Policy and Planning at Tufts University, Boston-Medford, MA. He is co-founder and co-editor of the international journal *Local Environment: The International Journal of Justice and Sustainability.* His more than 130 publications include *Local Environmental Policies and Strategies* (Longman, 1994), *Just Sustainabilities: Development in an Unequal World* (MIT Press, 2003), *Sustainable Communities and the Challenge of Environmental Justice* (NYU Press, 2005), *The New Countryside? Ethnicity, Nation and Exclusion in Contemporary Rural Britain* (The Policy Press, 2006), and *Environmental Justice and Sustainability in the Former Soviet Union* (MIT Press, 2009). He is a Fellow of the Royal Society of the Arts (FRSA), a member of the board of the Center for Whole Communities in Vermont, and is on the editorial boards of numerous journals. His research focuses on three areas: the nexus between the concepts of environmental justice and sustainability and, specifically, the possibility of a "just sustainability"; the extent, complexity, and pervasiveness of "rural racism" in Britain; and the potential role of "education for sustainability" in delivering more just and sustainable futures.

S. Harris Ali is an Associate Professor at the Faculty of Environmental Studies at York University, Toronto. His research interests involve the study of environmental health issues and the sociology of disasters and risk from an interdisciplinary perspective. He has published on toxic contamination events and disease outbreaks in journals including *Social Problems, Social Science and Medicine, Canadian Review of Sociology and Anthropology, Journal of Canadian Public Policy, Urban Studies,* and *Geography Compass.* He has recently co-edited (with Roger Keil) a volume entitled *Networked Disease: Emerging Infections in the Global City* (Wiley-Blackwell).

Jamie Baxter is an Associate Professor in the Department of Geography at the University of Western Ontario. He has been studying community perceptions of technological hazards, and hazardous facility siting for more than fifteen years. That work inevitably led to a more recent focus on environmental equity and justice. All of his major projects have been supported by the Social Sciences and Humanities Research Council of Canada; his current project is entitled *Environmental Inequity in Canada.*

F. Stuart Chapin III is an ecologist at the University of Alaska Fairbanks and a member of the National Academy of Sciences. He studies the impacts of climate change on Alaskan ecology and rural residents.

Peter Cole is a member of the Douglas First Nation (Stl'atl'imx) and Associate Professor in Aboriginal and Northern Studies, Faculty of Arts, at the University College of the North. His teaching and research interests include orality, narrativity, Aboriginal education, environmental thought, self-governance, and Aboriginalizing methodology. Peter has published in national and international academic and literary journals and is the author of the book *Coyote and Raven Go Canoeing: Coming Home to the Village* (McGill-Queen's University Press, 2006). He is currently the Principal Investigator of a Social Sciences and Humanities Research Council of Canada Standard Research Grant to work with the Stl'atl'imx on language and cultural regeneration. When he is not wearing an academic hat, Peter can be found hiking or canoeing.

Leith Deacon is a PhD candidate in the Department of Geography at the University of Western Ontario. His primary research interest lies in environmental equity and justice and the role of community perceptions. He also has interest in research methodology and hazards and disaster mitigation.

Lawrence K. "Larry" Duffy is President of the American Society of Circumpolar Health Board of Directors. Working in the areas of neurochemistry and toxicology for twenty years, Dr. Duffy has an extensive background in research and administration. He is currently studying the interaction of melatonin and toxins in brain development.

John Eyles is a Distinguished University Professor at McMaster University, in Hamilton, Ontario, where he is also Professor of Geography with cross-appointments in Sociology and Clinical Epidemiology and Biostatistics. He is author or editor of ten books and more than a hundred peer-reviewed papers on such topics as environmental and population health, risk management, and methodological development. He is also a fellow of the Royal Society of Canada.

Anna Godduhn is an interdisciplinary PhD candidate at the University of Alaska Fairbanks, studying military waste and nutrition in wild foods around Northway, Alaska. Her work includes perceptions of contamination and subsequent changes to food choice as potential risk factors of poor health outcomes.

A father of two and former wilderness guide, **Randolph Haluza-DeLay** is also an Assistant Professor of Sociology at The King's University College in Edmonton. His research interests have focused on environmental movements, anti-racism, environmental education, and religion and environmental issues. He is co-editor of a special issue of *Local Environment* on environmental justice in Canada and is currently involved in projects on the political ecology of Alberta and the cultural politics of the environment. He has also been employed as a youth counsellor, ski patroller, and director of an environmental education centre.

Lorelei L. Hanson is an Associate Professor of Environmental Studies and Human Geography at Athabasca University in Alberta. Her current research interests include rural community sustainability, Alberta identity and place, land trusts, and the use of mediated communication technologies in environmental education.

Henry P. Huntington is an independent research scientist from Eagle River, Alaska, and is one of the lead authors of the *Arctic Climate Impact Assessment*. His work addresses various ways that people interact with their environment, including subsistence, traditional knowledge, climate change, and conservation. He has worked in many communities in Alaska and throughout the Arctic.

Beenash Jafri is currently pursuing a PhD in the graduate program in Women's Studies at York University, Toronto. Her research interests include the politics of identity, anti-racism organizing, Indigenous solidarity and environmental justice. She was a founding member of the Anti-Racist Environmental Coalition and has been an educator and organizer for environmental justice in Toronto.

Roger Keil is the Director of the City Institute and Professor at the Faculty of Environmental Studies at York University, Toronto. He has published widely on urban and environmental politics and governance. Roger is the co-editor of the *International Journal of Urban and Regional Research* and a co-founder of the International Network for Urban Research and Action.

Gary Kofinas is Associate Professor of Resource Policy and Management in the Department of Resources Management and Institute of Arctic Biology at the University of Alaska Fairbanks (UAF). He is Director and Primary Investigator of the Resilience and Adaptation Program at UAF, an Integrative Graduate Education Traineeship of the National Science Foundation. His research focuses on sustainability of subsistence-based indigenous communities.

Bonita Lawrence (Mi'kmaq) teaches Indigenous Studies at York University, Toronto. She is the author of *"Real" Indians and Others: Mixed-Blood Urban Native People and Indigenous Nationhood* (UBC Press, 2004) as well as co-editor (with Kim Anderson) of *Strong Women Stories: Native Vision and Community Survival* (Sumach Press, 2003), and an edition of *Atlantis: A Women's Studies Journal,* entitled *Indigenous Women: The State of Our Nations*. A traditional singer, she has sung at rallies, social events, benefits, and prisons in Kingston and Toronto. She also currently volunteers with a diversion program for Aboriginal offenders at Aboriginal Legal Services of Toronto.

Robert Lovelace, retired Chief of Ardoch Algonquin First Nation, teaches at Queen's University and Fleming College in Ontario. Bob was born into a line of Tslagi Indians and was formally adopted into the Algonquin community in 1992 after years of partnership, including a legal defence of the wild rice stand near Ardoch. Bob has worked diligently to serve the Algonquin people in many facets of life, including politically, in court, in the prison system, as a counsellor, as a mediator, as a university professor, as acting chief, as co-chief, and by using tradition and ceremony in his daily life. He is an eloquent spokesman for Native rights, utilizing both English and Algonquin languages. Bob has four unique and spirited children at home, as well as three older children.

For over fifteen years **Deborah McGregor** has been an educator and trainer at both the university and community levels and has been involved in curriculum development, research, and teaching. Her focus is on indigenous knowledge in relation to the environment. She is an Assistant Professor at the University of Toronto, cross-appointed in Geography and Aboriginal Studies. She has taught various courses on indigenous knowledge as part of the Aboriginal Studies Program. *Miigwetch.*

David C. Natcher is an Associate Professor in the Department of Bioresource Policy and Academic Chair of the Indigenous Peoples Resource Management Program at the University of Saskatchewan, Saskatoon.

Melissa Ollevier graduated from York University, Toronto, with a master's in Environmental Studies in 2008. Her major paper examined the effectiveness of the Species at Risk Act in terms of protecting Canadian endangered species. She completed her Bachelor of Arts (honours) at the University of Toronto in 2006, with a double major in Environmental Studies and Criminology. She has conducted research for several environmental groups and government departments.

Bernard Ominayak has been Chief of the Lubicon Lake Indian Nation since 1978. The Lubicon people have been struggling to achieve recognition of their land rights and protection of their traditional territory for more than seventy years.

Pat O'Riley is an Associate Professor in the Department of Equity Studies, Faculty of Liberal Arts and Professional Studies, York University, Toronto. Her teaching and research interests are transdisciplinary: indigenizing research methodology, culturally relevant and environmentally sustainable discourses in education, poststructural theories and practices, and curriculum theory. Pat is the author of *Technology, Culture and Socioeconomics: A Rhizoanalysis of Educational Discourses* (Peter Lang Publishing, 2003). Pat is of Irish, Mohawk, and French heritage and married into the Stl'atl'imx in British Columbia, with whom she is conducting a research project on cultural and language regeneration. A proud mother and grandmother, Pat also loves canoeing and trekking.

Barbara Rahder immigrated to Canada from the United States in 1974. She received her PhD in Urban and Regional Planning from the University of Toronto in 1985. For more than thirty years, her research has explored the relationship between social and spatial processes, emphasizing issues of equity and access to affordable housing, community services, and public space. Barbara is Dean of the Faculty of Environmental Studies at York University in Toronto.

Maureen G.Reed is Professor in the School of Environment and Sustainability and the Department of Geography and Planning at the University of Saskatchewan, Saskatoon. Maureen is particularly concerned to explain social and equity dimension of environmental and land-use policies as they affect rural places; hence, her research is focused on how participatory decision-making approaches, working conditions, gender relations, and socio-cultural change affect the capacity of rural communities to work toward sustainability. Her book *Taking Stands: Gender and the Sustainability of Rural Communities* (UBC Press, 2003) was awarded the K.D. Srvastava Prize for academic excellence. When not doing research, Maureen enjoys family holidays across the country and admiring other people's gardens.

Kevin Thomas is an advisor to Chief and Council of the Lubicon Nation. He served as Chief Negotiator for the Lubicon Nation in land rights negotiations with the federal and provincial governments.

Sarah Fleisher Trainor is Research Assistant Professor at the University of Alaska Fairbanks. Her research focuses on climate change impacts, adaptation, and planning in Alaska, including specific attention to indigenous and scientific approaches and the science-policy interface. She is coordinator of the Alaska Center for Climate Assessment and Policy, one of several National Oceanic and Atmospheric Administration-funded Regional Integrated Sciences and Assessments programs and is Stakeholder Liaison for the Scenario Network for Alaska Planning, a University of Alaska capacity-building program for climate-change planning. She holds a PhD and a MA in Energy and Resources from the University of California, Berkeley.

Erica Tsang is a graduate of the Master in Environmental Studies program at York University, Toronto. Her interests in urban planning stem from her Bachelor of Environmental Studies from the University of Waterloo. Through her work, she inspires to harmonize sustainable built form and the land-development industry.

Index

Note: POPs stands for persistent organic pollutants;
TRCA, for Toronto and Region Conservation Authority

Printed and bound in Canada by Marquis Book Printing

Set in Stone by BookMatters, Berkeley

Copy-edited by Judy Phillips

Proofread by Sarah Munro

Indexed by Patricia Buchanan